STRIKING THE BALANCE BETWEEN NUTRIENT REMOVAL IN WASTEWATER TREATMENT AND SUSTAINABILITY

Michael W. Falk
HDR Engineering

JB Neethling
HDR Engineering

David J. Reardon
HDR Engineering

2011

WERF

The Water Environment Research Foundation, a not-for-profit organization, funds and manages water quality research for its subscribers through a diverse public-private partnership between municipal utilities, corporations, academia, industry, and the federal government. WERF subscribers include municipal and regional water and wastewater utilities, industrial corporations, environmental engineering firms, and others that share a commitment to cost-effective water quality solutions. WERF is dedicated to advancing science and technology addressing water quality issues as they impact water resources, the atmosphere, the lands, and quality of life.

For more information, contact:
Water Environment Research Foundation
635 Slaters Lane, Suite G-110
Alexandria, VA 22314-1177
Tel: (571) 384-2100
Fax: (703) 299-0742
www.werf.org
werf@werf.org

This report was co-published by the following organization.

IWA Publishing
Alliance House, 12 Caxton Street
London SW1H 0QS, United Kingdom
Tel: +44 (0) 20 7654 5500
Fax: +44 (0) 20 7654 5555
www.iwapublishing.com
publications@iwap.co.uk

Library of Congress Catalog Card Number: 2011933033

IWAP ISBN: 978-1-78040-043-3/1-78040-043-8

HDR Engineering

The research on which this report is based was developed, in part, by the United States Environmental Protection Agency (EPA) through Cooperative Agreement No. EM-83406901-0 with the Water Environment Research Foundation (WERF). However, the views expressed in this document are not necessarily those of the EPA and EPA does not endorse any products or commercial services mentioned in this publication. This report is a publication of WERF, not EPA. Funds awarded under the Cooperative Agreement cited above were not used for editorial services, reproduction, printing, or distribution.

ACKNOWLEDGMENTS

The project team would like to recognize Denver Metropolitan Wastewater Reclamation District (DMWRD), in particular Catherine Gerali and Barbara Biggs, for hosting the sustainability workshop in August 2010, and the presenters: Barbara Biggs (DMWRD), Dave Akers (Colorado DPHE), Phil Zahreddine (U.S. EPA), Robbin Finch (City of Boise) and Rod Reardon (Carollo). Several workshop attendees also graciously reviewed and provided comments on the first draft of the report: John Bratby (Brown and Caldwell); Art Umble (Greeley and Hansen); Shao-Yuan Leu (UCLA); Bob Brobst (U.S. EPA); Joe Husband (Malcolm Pirnie); Krishna Pagilla (IIT); and Tom Wilson (TEW Consulting).

Report Preparation

Principal Investigator:
JB Neethling, Ph.D., PE, BCEE
HDR Engineering

David J. Reardon, MS, BCEE, PE
HDR Engineering

Project Team:
Michael W. Falk, Ph.D., P.E.
HDR Engineering

Project Technical Review Committee

Kartik Chandran, Ph.D., PE
Columbia University

David J. Kinnear, Ph.D., PE
HDR Engineering

Michael K. Stenstrom, Ph.D., PE
University of California at Los Angeles

Issue Area Team (IAT) Members

James L. Barnard, Ph.D., D.Ing.h.c., Pr.Eng. BCEE
Black & Veatch

Joseph A. Husband, P.E., BCEE
Malcolm Pirnie

Gary R. Johnson, P.E., BCEE
Environmental Operating Solutions, Inc.

Sudhir Murthy, Ph.D., P.E.
DC Water

Tung Nguyen
Sydney Water Corporation

Clifford W. Randall, Ph.D.
Virginia Tech

G. David Waltrip, P.E.
Hampton Roads Sanitation District

James Wheeler, P.E.
U.S. Environmental Protection Agency

Kenneth N. Wood, P.E.
DuPont Company

Heng Zhang, Ph.D., P.E.
Metropolitan Water Reclamation Dist of Greater Chicago

Robbin W. Finch
Boise City Public Works

David Jenkins, Ph.D.
University of California – Berkeley

James A. Hanlon
U.S. Environmental Protection Agency

Terry L. Johnson, Ph.D., P.E., BCEE
Black & Veatch

Matthew Ries, P.E.
Water Environment Federation

Water Environment Research Foundation Staff

Director of Research: Daniel M. Woltering, Ph.D.
Senior Program Director: Amit Pramanik, Ph.D., BCEEM

ABSTRACT AND BENEFITS

This study focuses on sustainability impacts as wastewater treatment plants implement treatment technologies to meet increasingly stringent nutrient limits. The objective is to determine if a point of "diminishing returns" is reached where the sustainability impacts of increased levels of nutrient removal outweigh the benefits of better water quality.

Five different hypothetical treatment trains at a nominal 10 mgd flow were developed to meet treatment targets that ranged from cBOD mode (Level 1) to four different nutrient removal targets. The nutrient removal targets ranged from 8 mg N/L; 1 mg P/L (Level 2) to the most stringent at <2 mg N/L; <0.02 mg P/L (Level 5).

Given that sustainability is a broad term, the industry-accepted three pillars of sustainability were evaluated and discussed, and particular emphasis was placed on the environmental and economic pillars. The following variables received the most attention: greenhouse gas (GHG) emissions, a water quality surrogate that reflects potential algal growth, capital and operational costs, energy demand, and consumables such as chemicals, gas, diesel, etc.

The results from the GHG emissions metric are shown below. Note that biogas cogeneration is represented by negative values as biogas production can be used to offset energy demands. The nitrous oxide (N_2O) emissions values are based on the average biological nutrient removal (BNR) and non-BNR plants evaluated in the United States national survey by Ahn et al. (2010b). The error bars represent the data range of the national survey.

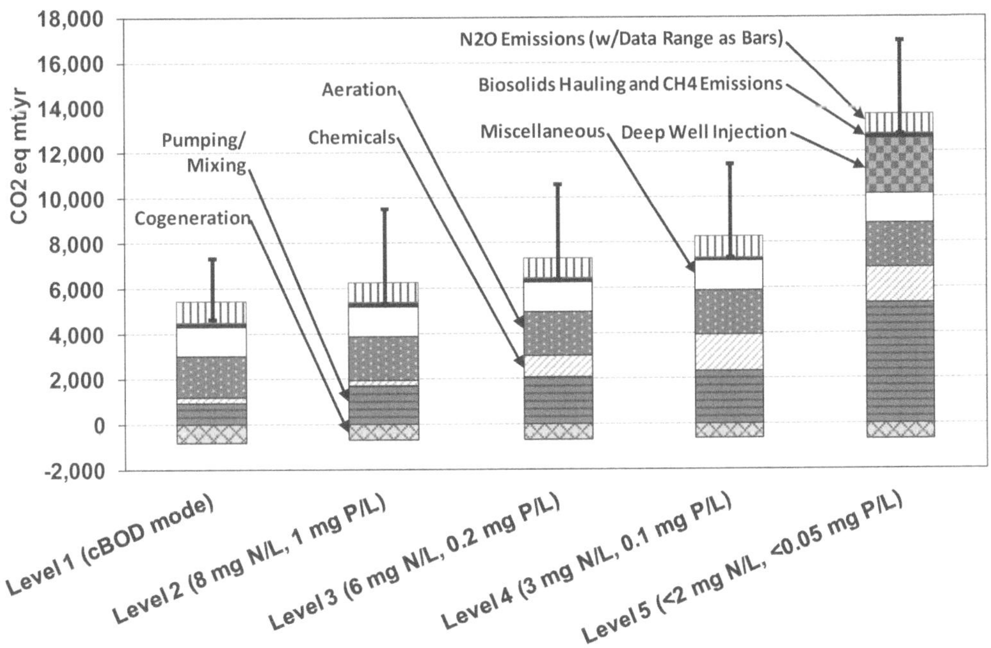

The GHG emissions results suggest that a point of diminishing return is reached at Level 4 (3 mg N/L; 0.1 mg P/L). The GHG emissions show a steady increase from Levels 1 to 4, followed by a 65% increase when moving from Level 4 to 5. Despite a 70% increase in GHGs, the discharged nutrient load only decreases by 1% by going from Level 4 to 5. The primary contributors to GHG emissions are energy related (aeration, pumping, mixing). The GHG emissions associated with chemical use increases for the more stringent nutrient targets that required chemical treatment in addition to biological nutrient removal.

In terms of cost, the total project capital cost increases approximately one-third from $9.3 million to $12.7 million for changing from Level 1 to 2, followed by a more than doubling in cost when changing from Level 1 to 5. Total project capital costs in this report are for a Greenfield plant. The operational cost increase between levels is more pronounced than total project capital cost with more than five-times increase from Level 1 to 5 ($250/MG treated to $1,370/MG treated, respectively).

Although the GHG emissions and cost of wastewater treatment increases as nutrient removal is implemented, it is important to note some benefits associated with plants performing nutrient removal:

- Lower BOD/TSS discharge load
- Higher removal of trace organic compounds (TrOCs) and heavy metals
- Improved water quality and benefit to downstream users
- Less algal growth
- Reduced receiving water dissolved oxygen demand due to ammonia removal
- Reduced ammonia discharge loads, which is toxic to several aquatic species
- Improved water quality for habitat, especially as it relates to biodiversity and eutrophication
- Secondary clarifier effluent more conditioned for filtration and disinfection
- Greater process stability from the anaerobic/anoxic zones serving as a selector

This report focused on in-plant (point source) options for nutrient removal and the implications for cost and sustainability. Other approaches, such as addressing non-point sources, could be added to the assessment. Rather than focusing strictly on point source dischargers and requiring Level 4 or 5 treatments, Level 3 or 4 treatments complimented with best management practices of non-point sources might be a more sustainable approach at achieving comparable water quality.

Benefits

- Highlights sustainability impacts at higher levels of nutrient removal, and how these impacts might outweigh benefits of reduced nutrient discharge loads.
- Provides an approach for GHG emissions analysis that considers factors outside a municipality's boundary, such as hauling and chemical manufacturing.

- Quantifies the differences in sustainability between a secondary and an advanced nutrient removal wastewater treatment plant (WWTP).
- Highlights the distribution of GHG emissions contributors within a wide range of nutrient removal objectives.
- Lists the ancillary benefits of a longer SRT associated with nitrification plants.

Keywords: Greenhouse gas emissions, wastewater treatment plants, sustainability, nutrient removal, triple bottom line.

TABLE OF CONTENTS

LIST OF TABLES

LIST OF FIGURES

LIST OF SYMBOLS AND ABBREVIATIONS

3-stage Phoredox	A2O
Advanced Oxidation Processes	AOPs
Ammonia	NH_4
Annual Average	AA
Best Management Practices	BMPs
Biochemical Oxygen Demand	BOD
Biological Nutrient Removal	BNR
British Thermal Units	BTU
Carbon Dioxide	CO_2
Carbon Dioxide Equivalent Metric Tonnes per Year	CO_2 eq mt/yr
Carbonaceous Biochemical Oxygen Demand	cBOD
Cubic Feet	cf
Day	d
Deep Well Injection	DWI
Dimethylamine	DMA
Dissolved Combined Amino Acids	DCAA
Dissolved Free Amino Acids	DFAA
Dissolved Organic Nitrogen	DON
Dissolved Organic Phosphorus	DOP
Dissolved Organic Phosphorus	DOP
Dissolved Oxygen	DO
Ethylenediaminetetraacetic acid	EDTA
Gallon	gal
Gallons per Square Foot per Day	gfd
Gallons per Minute	gpm
Gigawatt Hours	GWh
Gram	g
Gravity Belt Thickener	GBT
Greenhouse Gas	GHG
Heating Ventilation Air-Conditioning	HVAC
High Rate Clarification	HRC
Horsepower	hp
Hour	hr

Hydraulic Residence Time	HRT
International Panel on Climate Change	IPCC
Kilogram	kg
Kilowatt Hours	kWh
Life-Cycle Analysis	LCA
Mean Particle Size	MPS
Megawatt Hours	MWh
Methane	CH_4
Microfiltration	MF
Million British Thermal Units	MMBTU
Million Gallons per Day	mgd
Mixed Liquor Return	MLR
Mixed Liquor Suspended Solids	MLSS
Modified Ludzack-Ettinger	MLE
Multi-Effect Distillation	MEDN
Nanofiltration	NF
National Association of Clean Water Agencies	NACWA
National Pollution Discharge Elimination System	NPDES
Natural Resources Defense Council	NRDC
Nitrate	NO_3
Nitric Oxide	NO
Nitrification and Denitrification	NDN
Nitrilotriacetic Acid	NTA
Nitrogen	N
Nitrous Oxide	N_2O
Particulate Acid Hydrolysable Phosphorus	pAHP
Particulate Organic Phosphorus	pOP
Particulate Reactive Phosphorus	pRP
Phosphate	PO4
Phosphorus	P
Phosphorus Accumulating Organisms	PAOs
Population Equivalent	PE
Pound	lb
Pounds per Day per Square Foot	lb/d/sf
Pounds per Square Inch	psi
Refractory Dissolved Organic Nitrogen	rDON
Refractory Dissolved Organic Phosphorus	rDOP

Return Side Streams	RSS
Reverse Osmosis	RO
Salinity Gradient Solar Pond	SGSP
Sequencing Batch Integrated Fixed-Film Activated Sludge	SB-IFAS
Solids Retention Time	SRT
Soluble Acid Hydrolysable Phosphorus	sAHP
soluble BOD	sBOD
Soluble Reactive Phosphorus	sRP
Square Feet	sf
Struvite	$MgNH_4PO_4 \cdot 6H_2O$
Sustainable Return on Investment	SROI
Thousand British Thermal Units	kBTU
Total Dissolved Solids	TDS
Total Kjeldahl Nitrogen	TKN
Total Maximum Daily Load	TMDL
Total Nitrogen	TN
Total Phosphorus	TP
Total Suspended Solids	TSS
Trace Organic Compounds	TOrCs
Triple Bottom Line	TBL
Ultimate Biochemical Oxygen Demand	uBOD
United States Environmental Protection Agency	U.S. EPA
Wastewater Treatment Plant	WWTP
Year	yr

EXECUTIVE SUMMARY

ES.1 Background and Goals

The United States Environmental Protection Agency (U.S. EPA) has determined that high nitrogen (N) and phosphorus (P) loadings result in harmful algal blooms that reduce spawning grounds and habitat, fish kills, oxygen-starved hypoxic or "dead" zones, public health concerns related to impaired drinking water sources, and increased exposure to toxic microbes (U.S. EPA, 2010).

Four important watersheds that greatly concern the U.S. EPA are:

- Chesapeake Bay. This watershed has an existing hypoxia problem and 150,000 new people are added each year to the watershed.
- The Gulf of Mexico. This watershed has a prevalent and well-documented hypoxic "dead" zone. There are 31 states that contribute to the watershed. Nitrogen is the focus of efforts to reduce the hypoxia zone, but the 2001 Hypoxia Action Plan of the U.S. EPA Science Advisory Board reports that phosphorus plays a much greater role in the hypoxia problem than previously thought.
- Long Island Sound. Dissolved oxygen (DO) is below standards in one-third to one-half of the Sound. Nitrogen loadings have been capped at 1990 loads. A water quality trading program has been implemented in Connecticut for point sources with a market-based approach.
- Puget Sound. The Puget Sound Action Plan calls for a better understanding of nutrient and bacteria loadings from septic systems.

The U.S. EPA's approach toward issuing nutrient discharge limits is based on site-specific standards. The U.S. EPA uses a concept called "ecoregions" which represent areas of similar climate, hydrology, geology, physiography, soils, land use, vegetation and wildlife (U.S. EPA, 2000). Other government agencies and grassroots organizations contend that nutrient pollution is widespread and justifies a general standard.

The Natural Resources Defense Council (NRDC) has petitioned the U.S. EPA to develop a quantified nutrient pollution discharge limit for wastewater treatment plants (WWTPs). The Sierra Club, American Rivers, and others have joined NRDC in this petition. NRDC argues that the U.S. EPA must redefine secondary treatment to include nutrient removal. NRDC has recommended that the following effluent nutrient levels be established:

- Effluent total phosphorus (TP) at 1.0 mg P/L and total nitrogen (TN) at 8.0 mg N/L, which are consistently attainable using current biological based technologies.
- Effluent TP at 0.5 mg P/L and TN at 5 mg N/L, which are consistently attainable using current biological based technologies that might require complementary chemicals (e.g., external carbon source for denitrification).
- Effluent TP at 0.3 mg P/L and TN at 3.0 mg N/L, which is attainable with existing technology using only improved biological and chemical treatment processes.

The discussion to restrict nutrient discharges, namely nitrogen and phosphorus, has largely focused on receiving water quality and has ignored the corresponding impact on sustainability from treatment. This report presents a bench-top analysis on the balance between nutrient removal and sustainability.

Sustainability means different things to different people. The classic definition of sustainability was created by the Brutland Commission, formally the World Commission on Energy Development (1987):

"Sustainable development is development that meets the needs of the present without compromising the ability of future generations to meet their own needs."

This general description serves as a starting point. However, quantifiable metrics are needed to determine and assess the key components of sustainability. The three areas typically referenced when performing sustainability evaluations are the environmental, economic and social pillars. These pillars are commonly referred to as the triple bottom line (TBL), as illustrated in Figure 1-1. The TBL is based on the notion that all three pillars are incorporated into the decision-making process to promote a sustainable solution.

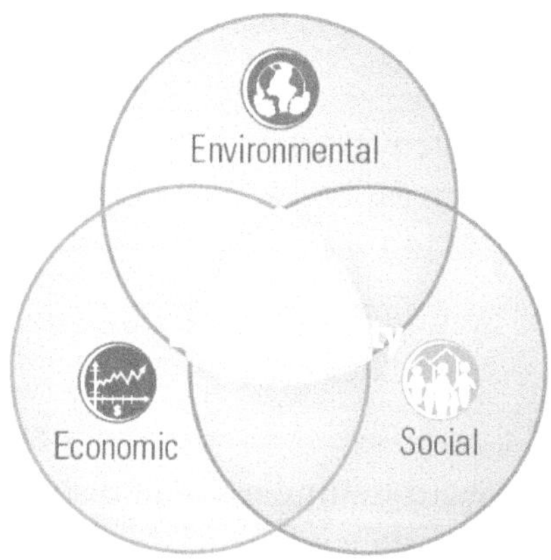

Figure ES-1. Triple Bottom Line Schematic to Achieve Sustainability.

For this investigation, the researchers performed analyses on the environmental and economic components of the TBL. The environmental component considered GHG emissions and a water quality surrogate, potential algal production. For the economic pillar, capital and operations costs were evaluated. Because the social pillar is case specific, we did not carry out any analyses. Instead, the social pillar is discussed in Chapter 4.0 but not quantified. Overall, this is a simplified approach instead of a WWTP specific methodology that would investigate an array of economic, social and environmental sustainability impacts.

The objective of this study is to determine if a point of diminishing returns is reached where the sustainability impacts of increased levels of nutrient removal outweigh the benefits of improved water quality. Increased complexity and cost of nutrient removal (water quality goal) may also lead to increased environmental impacts in terms of GHG emissions (encompasses energy, manufacturing, hauling, air pollution, etc.). Given that sustainability is a broad term, the focus was placed on the following variables while developing the comparative model:

- GHG emissions (related to energy demand and potential air quality impacts)
- Capital costs
- Operating costs
- Energy demand
- Water quality
- Consumables, such as chemicals, gas, diesel, etc.

ES.2 Five Levels of Treatment

The analysis considered the sustainability impacts of treatment plants designed to achieve five different levels of nutrient removal as shown in Table ES-1. With each treatment level, a process analysis was developed to quantitatively determine the GHG emissions equivalents based on the aforementioned parameters. The treatment process trains are outlined in Table ES-2.

Table ES-1. Treatment Level Objectives.

Level	Biochemical Oxygen Demand (mg/L)	Total Suspended Solids (mg/L)	Total Nitrogen (mg N/L)	Total Phosphorus (mg P/L)
1	30	30	-	-
2	<30	<30	8	1
3	<30	<30	4-8	0.1-0.3
4	<30	<30	3	0.1
5	<30	<30	<2	<0.02

Table ES-2. Unit Processes per Treatment Level.

Level	Liquid Treatment	Solids Treatment	Comment
1	Primary Clarifier Activated Sludge Disinfection Dechlorination	Gravity Belt Thickener Anaerobic Digestion with Cogen Centrifugation	Conventional Activated Sludge for BOD/TSS removal
2	Primary Clarifier Activated Sludge Alum (optional) Disinfection Dechlorination	Gravity Belt Thickener Anaerobic Digestion with Cogen Centrifugation	Nitrification/Denitrification and Biological Phosphorus Removal
3	Primary Clarifier Activated Sludge Methanol (optional) Alum (filtration) Filtration Disinfection Dechlorination	Gravity Belt Thickener Anaerobic Digestion with Cogen Centrifugation	Nitrification/Denitrification and Biological Phosphorus Removal and Filtration
4	Primary Clarifier Activated Sludge Methanol (optional) Alum/Polymer (Enhanced Settling) Enhanced Settling Filtration Disinfection Dechlorination	Fermentation Gravity Belt Thickener Anaerobic Digestion with Cogen Centrifugation	Nitrification/Denitrification and Biological Phosphorus Removal, High Rate Clarification and Denitrification Filtration
5	Primary Clarifier Activated Sludge Methanol (optional) Alum/Polymer (Enhanced Settling) Enhanced Settling Filtration Microfiltration Reverse Osmosis Disinfection Dechlorination	Gravity Belt Thickener Anaerobic Digestion with Cogen Centrifugation	Nitrification/Denitrification and Biological Phosphorus Removal, High Rate Clarification, Denitrification Filtration, and MF/RO on about Half the Flow

ES.3 Results

The investigation determined that as the degree of nutrient removal increased, the impact on sustainability increased significantly. In particular, the GHG emissions and cost increased significantly. Results are presented in Table ES-3 and Figure ES-2. For the purpose of consistency, all plants were evaluated at a nominal flow of 10 mgd.

Note that the negative GHG emissions values represent the benefit of anaerobic digestion with cogeneration (resulting in a reduced amount of electricity purchased), which is assumed for all levels of treatment. The nitrous oxide (N_2O) emissions are based on the average values developed in the U.S. nationwide survey for both secondary treatment and nutrient removal WWTPs (Ahn et al., 2010b). The error bars represent the data range from the U.S. national survey.

The amount of GHG emissions increases 20% as the treatment objective increases from Level 1 to 2. The GHG emissions increase roughly 15% for each step from Level 2 to 3 and Level 3 to 4 (30% from Level 2 to 4). The increase is primarily due to additional chemicals (methanol, alum, and polymer) and energy demand (filtration).

Achieving treatment Level 5 (TN<2 mg N/L and TP<0.02 mg P/L) resulted in a dramatic increase in GHG emissions (presented as CO_2 equivalent metric tonnes per year (CO_2 eq mt/yr)) with a 70% increase of GHG emissions from treatment Level 4 alone while the discharged nutrient load only decreases by 1% by going from Level 4 to 5. This sharp increase in GHG emissions and small benefit in nutrient reduction suggests that a point of diminishing returns may be reached since a small improvement in water quality caused a very large impact on the overall environment.

Table ES-3. Greenhouse Gas Emission Results for Five Levels of Nutrient Treatment (10 mgd).

Treatment	Level 1 No N/P Removal (CO_2 eq mt/yr)	Level 2 8 mg N/L 1 mg P/L (CO_2 eq mt/yr)	Level 3 4-8 mg N/L 0.1-0.3 mg P/L (CO_2 eq mt/yr)	Level 4 3 mg N/L 0.1 mg P/L (CO_2 eq mt/yr)	Level 5 2 mg N/L <0.02 mg P/L (CO_2 eq mt/yr)
Primary	240	240	240	240	240
Activated Sludge	2,980	3,820	4,430	4,430	4,430
Tertiary	200	200	650	1,530	6,990
Sub-Total Liquid Stream	3,420	4,260	5,320	6,200	11,660
Solids *	-150	-10	-40	50	-30
Miscellaneous **	1,320	1,320	1,320	1,320	1,320
Total	**4,590**	**5,570**	**6,600**	**7,570**	**12,950**

* The solids treatment stream includes cogeneration of biogas, which results in negative values for most treatment levels

** Miscellaneous results include odor control, non-potable water, high pressure air system, miscellaneous pumping/mixing/sludge storage/transfer pumps, HVAC, administrative buildings energy, and miscellaneous equipment.

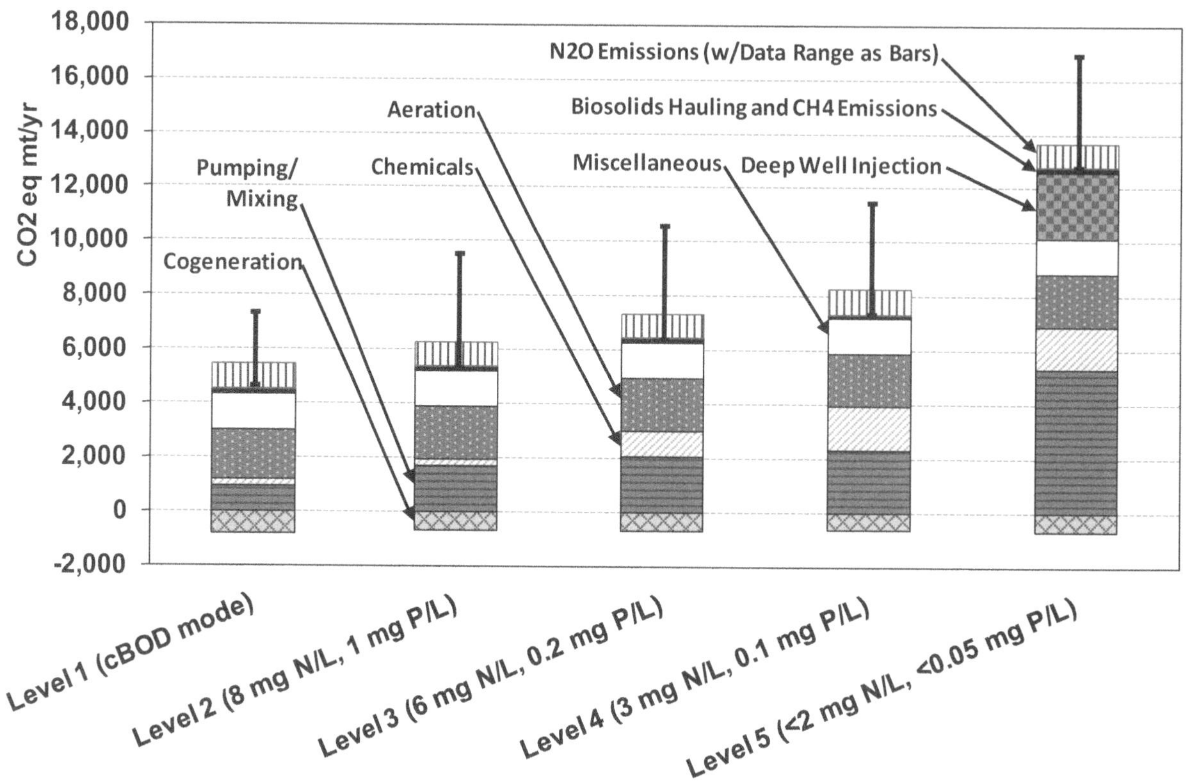

Figure ES-2. Combined Liquid and Solids Stream Mass Balance GHG Emissions Results.
(10 mgd; 100,000 PE for 100 gal/d/capita)

The largest contributors to GHG emissions are energy, deep well injection, miscellaneous, chemicals, and nitrous oxide (N_2O) emissions. The primary contributor to GHG emissions is energy (pumping, mixing and aeration), followed by deep well injection (DWI). Given that brine management can be achieved by other approaches, the DWI value can be substituted with a different technology. The impact from chemicals increases as the treatment levels increase because chemicals are typically required to achieve more stringent effluent limits. The N_2O emissions are highly variable as indicated by the error bars that represent the data range from the database used to develop the values by Ahn et al. (2010b).

The smallest contribution came from fugitive methane (CH_4) emissions, which was calculated using the protocol developed by the IPCC (2006). Although the fugitive methane emissions calculations is in accordance with the IPCC protocol, the values are most likely understated as it only considers fugitive emissions during anaerobic digestion. Methane gases can form at other locations within the WWTP (e.g., headworks, primary clarifiers, etc.).

The above results are further magnified when the GHG emissions impacts are expressed in terms of the incremental amount of N and P removed by each level of treatment. As shown in Figure ES-3 on a log scale, the increase in GHG emissions as treatment level increases per additional pound removed of N increases exponentially beyond Level 4. As for P, the increase in GHG emissions with treatment per additional pound of P removed increases exponentially after Level 3. The figure indicates that the GHG emissions impact on the incremental removal (the last pound removed) is significantly higher than the initial benefit.

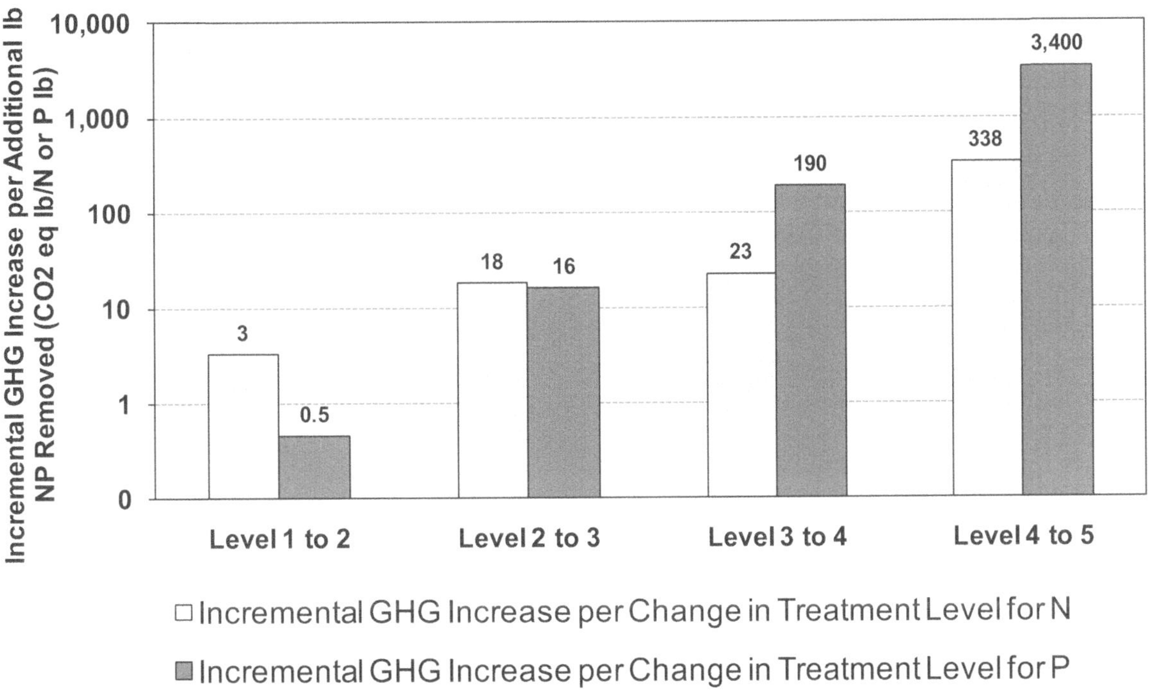

Figure ES-3. Incremental GHG Emissions Increase per Additional Pound N or P Removed.

Because the rationale behind nutrient removal is to improve water quality, the benefit of nutrient removal should be expressed in terms of reducing algal production. A surrogate parameter, the "potential algal production," was used as a means to assess water quality improvement. The molecular formula for algal composition determined by Vallentyne (1974) shows that every 100 pounds of algae produced requires 7.2 lb of N and 1.0 lb of P. This relationship can be used to calculate the reduction in algae production as a function of nitrogen or phosphorus reduction. The plot of GHG emissions along with algae production for the five treatment levels is shown in Figure ES-4 for the 10 mgd plant. Nearly 95% of the potential algae production is reduced by changing from Level 1 to Level 3 treatment. A remaining 4% of potential algae production is reduced from Levels 3 to 5, but at a nearly doubling of GHG emissions (6,590 to 12,950 CO_2 equivalent metric tonnes per year).

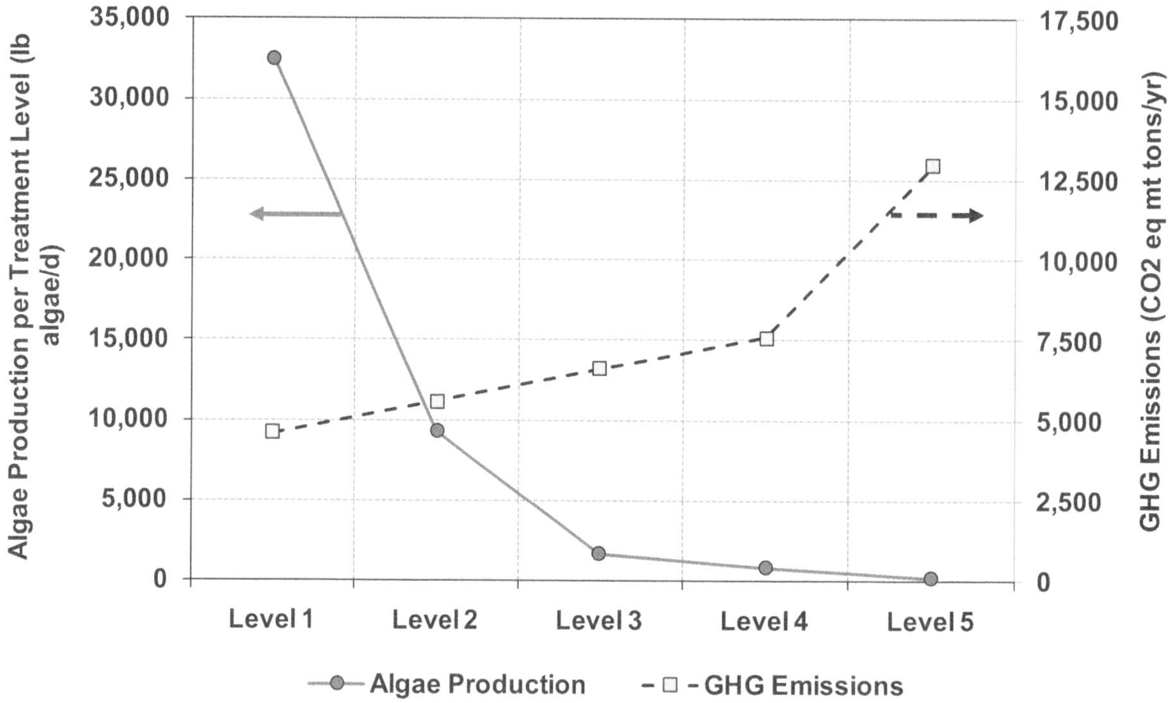

Figure ES-4. GHG Emissions and Algae Production per Treatment Level.

Although the increased GHG emissions values and reduced algal production potential, it is critical to note some ancillary benefits associated with transitioning from secondary treatment to nutrient removal:

- Lower BOD/TSS discharge load
- Higher removal of trace organic compounds (TrOCs) (Horz et al., 2004)
- Higher removal of heavy metals (Stasinakis et al., 2003)
- Improved water quality and benefit to downstream users
- Less algal growth as there is less nutrient load discharged
- Reduced receiving water dissolved oxygen demand due to ammonia removal
- Reduced ammonia discharge loads, which is toxic to several aquatic species
- Improved water quality for habitat, especially as it relates to biodiversity and eutrophication (Beck et al., in press)
- Secondary clarifier effluent more conditioned for filtration and disinfection (Leu et al., In press; Chan et al.; in press)
- Greater process stability from the anaerobic/anoxic zones serving as a selectors (Jenkins et al., 2004)

Operational sustainability impacts as measured by the aggregated GHG emissions (including pumping, mixing, aeration, chemicals manufacturing/hauling, and biosolids hauling) are not the only impact of increasingly restrictive nutrient limits. The cost also increases with more advanced treatment.

The impacts of the treatment levels on total project costs are shown in Table ES-4. It is important to highlight that total project capital cost are for a Greenfield plant. Increased treatment levels can more than double the total project capital cost compared to conventional secondary treatment. The difference in operations cost between treatment levels is more pronounced than total project capital cost as a direct result of increased pumping, mixing, chemicals and others required to achieve the more stringent levels. Total present worth cost were developed to illustrate the difference in treatment levels as both capital and operations should both be considered while doing a comparison.

Table ES-4. Total Project Capital and Operations Costs for a 10 mgd WWTP. [i]

Level	Total Project Capital Cost ($/gpd) [ii]	Total Project Capital Cost (Million $) [ii]	Operations Cost ($/MG Treated) [iii]	Operations Cost ($1,000/yr) [iii]	Total Present Worth Project Costs (Million $) [iv]
1 (No N/P Removal)	9.3	93	250	910	110
2 (8 mg N/L; 1 mg P/L)	12.7	127	350	1,260	150
3 (4-8 mg N/L; 0.1-0.3 mg P/L)	14.4	144	640	2,350	180
4 (3 mg N/L; <0.1 mg P/L)	15.3	153	880	3,200	210
5 (1 mg N/L; <0.02 mg P/L)	21.8	218	1,370	4,990 **	300

Notes:
i The total project capital cost are for a Greenfield plant.
ii The total project capital cost are the equipment cost, construction, and "soft costs"
iii Operations cost = energy and chemical cost. Labor and maintenance costs are excluded
iv The assumed discount rate was 5 percent at an escalation rate of 3.5 percent (capital, energy, non-energy)

In summary, increased levels of treatment for nutrients substantially impacts the sustainability of WWTPs in terms of GHG emissions and costs. The GHG emissions, total project capital, operations, and total present worth costs for high levels of nutrient removal increase rapidly with increased nutrient removal.

ES.4 Conclusions

1. Removing N and P to treatment Levels 4 and 5 results in negative sustainability impacts with a corresponding small improvement to water quality.
2. GHG emissions increase significantly when nutrient removal is incorporated in a wastewater treatment plant. For a nominal 10 mgd WWTP, the GHG emissions increased from 4,590 CO_2 equivalent metric tons per year for a cBOD mode WWTP (Level 1) to 7,570 CO_2 equivalent metric tons per year, where nitrogen and phosphorus removal requirements are 3 mg N/L and 0.1mg P/L (Level 4). Further reductions in N and P to <2 mg N/L and <0.02 mg P/L (Level 5) increase the GHG emissions to 12,950 CO_2 equivalent metric tons per year.
3. The level of nutrient removal also increases the total project capital, operating, and total present worth costs. The total project capital cost increased from $93 million for a Level 1 WWTP to $153 million for a Level 4 WWTP and $218 million for a Level 5 WWTP. The difference in operating cost between treatment levels is more pronounced than total project capital cost (Level 1 are $910,000 per year and $3,200,000 per year for a Level 4; more than

three times as much). The incorporation of total project capital and operating costs into a total present worth cost illustrates the overall difference in cost between treatment levels, where Level 4 is nearly twice as much as Level 1.

4. The incremental GHG emissions increase in lbs CO_2 eq per lb nutrient removed increases exponentially as the level of treatment increases. The incremental values from Level 1 to Level 2 are 3 for N and 0.5 for P. The intensity for moving from Levels 4 to 5 is 338 for N and 3,400 for P.
5. Recalcitrant dissolved organic nitrogen, commonly referred to as refractory dissolved organic nitrogen (rDON), impairs a WWTPs ability to reliably achieve low TN objectives. Effluent limits that require nitrogen values of 2 mg N/L or less might require the use of expensive and energy intensive strategies, such as reverse osmosis, that result in elevated GHG emissions.
6. As with rDON, recalcitrant dissolved organic phosphorus, commonly referred to as rDOP, hinders a WWTPs ability to meet the total phosphorus (TP) value required of Level 5. As a result, an energy intensive technology, such as reverse osmosis (RO), is required to ensure reliably meeting the treatment objective.
7. The GHG emissions for all treatment levels are dominated by energy demand.
8. Based on the incremental GHG emissions increase per additional pound nutrient removed, a point of diminishing returns appears to be reached between Levels 3 and 4 for nitrogen removal, and between Levels 4 and 5 for phosphorus removal.
9. Using reverse osmosis to achieve extremely low levels of nitrogen and phosphorus increases costs and GHG emissions. Brine reject management remains a challenge for reverse osmosis applications, especially for inland applications.
10. Ancillary benefits associated with a longer sludge age in nutrient removal treatment levels exist and should be accounted for in a full evaluation (e.g., secondary clarifier effluent more conditioned for filtration/disinfection, less biosolids produced, etc.).

ES.5 Future Work

1. Developing a rational approach to receiving water nutrient loading, such as requiring "best practices" for non-point sources, may be much more beneficial than requiring WWTPs to remove nutrients to levels that adversely impact sustainability. Further dialog with regulators on a national and local level is needed to find a regulatory framework that best protects water quality and manages GHG emissions using both non-point and point source control.
2. Clearly define whether N or P is the limiting nutrient for specific receiving waters.
3. Incorporate bioavailability of N and P in models and assessments to set permit limits. Consider the impact of rDON and rDOP and the slow transformation kinetics of these compounds in nature.
4. Determine the role of peaking factors in GHG emissions per capita for the different treatment levels, particularly in combined sewer plants (majority in United States). Oversized and underutilized facilities translate to higher GHG emissions per capita with increased levels of treatment.

5. Continue work on finding more sustainable technologies for removing nitrogen and phosphorus. Examples include using anammox bacteria to remove ammonia and recycling chemical sludge for phosphorus removal.
6. Initiate discussion with the U.S. EPA regarding the impact of refractory nitrogen and phosphorus on water quality and impact on treatment costs and sustainability.
7. Determine the role of coagulant chemical selection in removing refractory N and P.

CHAPTER 1.0

WASTEWATER TREATMENT PLANTS IMPACT ON SUSTAINABILITY

1.1 Sustainability

Sustainability means different things to different people. The classic definition of sustainability was created by the Brutland Commission, formally the World Commission on Energy Development (1987):

"Sustainable development is development that meets the needs of the present without compromising the ability of future generations to meet their own needs."

This general description serves as a starting point. However, quantifiable metrics are needed to determine and assess the key components of sustainability. The three areas typically referenced when performing sustainability evaluations are the environmental, economic and social pillars. These pillars are commonly referred to as the triple bottom line (TBL), as illustrated in Figure 1-1. The TBL is based on the notion that all three pillars are incorporated into the decision-making process to promote a sustainable solution.

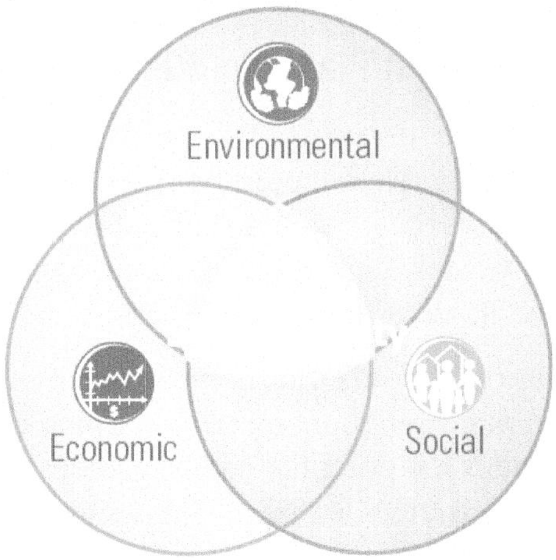

Figure 1-1. Triple Bottom Line Schematic to Achieve Sustainability.

Triple bottom line of sustainability includes satisfying or meeting economic, environmental, and social criteria to obtain the most sustainable solution. Some of the considerations used for defining sustainability in the wastewater world include, but are not limited to:

- Water quality and quantity (Rogers et al., 2002)
- Climate change using GHG emissions as a surrogate (Zimmerman et al., 2008)

- Consumables, such as chemicals (Kang et al., 2009)
- Discharge of emerging contaminants (Kolpin et al., 2002)
- Biodiversity reduction in receiving waterbodies (Beck et al., in Press)
- Energy, including natural gas, electricity and fuel (Oki et al., 2006)
- Waste, including solid waste and biosolids (NBP, 2007)
- Generation of potential resources, including biogas (Oki et al., 2006), recycled water (Zimmerman et al., 2008), biosolids (NBP, 2007), fertilizer (Larsen et al., 2009) and biodiesel (Conley et al., 2009)
- Economics, including lifecycle cost analysis (Lundie et al., 2004)
- Social and community impacts (Ashley et al., 2008)

A case study from the city of Boise, Idaho, which performed a sustainability evaluation using the TBL, was presented at workshop of the Annual Pacific Northwest Clean Water Association Conference (Finch, 2009). For this investigation, the researchers performed analyses on the environmental and economic components of the TBL. Within environmental, they considered GHG emissions and a water quality surrogate, potential algal production. For the economic pillar, capital and operations costs were evaluated.

Because the social pillar is case specific, the researchers did not carry out any analyses. Instead, the social pillar is discussed but not quantified. Social values are site-specific and are best quantified/estimated by the community that is considering the sustainability analysis.

Overall, this is a simplified approach instead of a facilities-based methodology that would investigate an array of economic, social and environmental sustainability impacts.

1.2 Role of Wastewater Treatment Plants in Sustainability

As indicated in the previous section, the TBL serves as a guide when performing sustainability analyses. Wastewater by itself should be considered a resource because it contains vital and potentially recoverable resources – water, nutrients, and energy (Guest et al., 2009). Historically, sustainability in the wastewater sector has focused on water quantity stressors (Oki and Kanae, 2006; Zimmerman et al., 2008; Guest et al., 2009).

With increases in population, urbanization, climate change, demographic, and land use changes, water stressors became more pronounced (Zimmerman et al., 2008; Guest et al., 2009). Focusing on water quantity excludes the potential removal/recovery of valuable nutrients. The wastewater sector needs to expand its view and incorporate water quality when addressing sustainability. A few positive impacts from nutrient removal at WWTPs on the three pillars include the following (Conley et al., 2009):

- Improved water quality for habitat, especially as it relates to biodiversity (Beck et al., 2009) and eutrophication.
- Improved water quality for downstream users increase water use flexibility. This can lead to more industrial jobs and benefits to agricultural on both the local and global levels.
- Reduced potential for impacts on receiving water dissolved oxygen.

- Reduced levels of ammonia, which is toxic to some aquatic species.

This list will be expanded as the impact of processes performing nutrient removal is incorporated below.

1.3 Measuring Sustainability

The statistical approaches to measuring sustainability are based on metrics that typically fail to simultaneously incorporate all three TBL pillars. The most advanced statistics exist for economics and environmental. Two commonly used techniques that incorporate economics/environmental are life-cycle analysis (LCA) (Foley et al., 2010) and sustainable return on investment (SROI) (Williams and Larocque, 2009). Of those, LCA is the most widely used tool to quantify environment/economic components (Guest et al., 2009).

LCA attempts to quantify environmental impacts of a product throughout its life-cycle, which is commonly referred to as "cradle to grave." Although LCA is proven in terms of incorporating environmental and economic components, it struggles as with most statistical means to incorporate a social component (Hunkeler and Rebitzer, 2005). Most available social metrics only consider public safety risk assessment (Hunkeler et al., 2006). Social metrics are not readily available that include non-risk aspects common to TBL analysis at WWTPs.

1.4 Energy Demand at Wastewater Treatment Plants

A nationwide breakdown found waste to represent 2.3% of the estimated US GHG emissions (U.S. EPA, 2007). Waste in this case entails landfills, wastewater treatment and composting. Landfills represent over 75% of the 2.3%, and wastewater treatment plants at less than 1% (i.e. less than 0.023 of the U.S. GHG emissions).

The largest contributor to GHG emissions from WWTPs is from energy demand (Keller and Hartley, 2003). WWTP power needs represent about 75% of the plant GHG emissions footprint (Scope 2 emissions) (Pagilla et al., 2009). Thus, Scope 2 emissions are the low-hanging fruit to reduce the GHG emissions footprint of a WWTP. For example, GHG emissions savings audits for municipalities that reduced energy demands can represent significant GHG emission savings opportunities on a local level.

Wett and others (2007) have shown that WWTPs can be designed to achieve zero net energy demand under controlled conditions. The WWTP in Strass, Austria, where Wett and his colleagues performed their research, had a permit (14 mg N/L total nitrogen) that lends itself towards carbon neutral operation. As the nitrogen discharge requirement drops below 14 mg N/L, the energy demand increases to a level where carbon neutrality is more difficult to achieve.

Given that the nitrogen removal levels considered in this paper are all less than 14 mg N/L, achieving carbon neutrality is highly unlikely. Additionally, North American plants receive a more dilute wastewater stream than European plants. Therefore, American plants have more difficulty in capturing carbon.

1.5 Cap and Trade

Greenhouse gas emissions occur directly from wastewater treatment plant processes, and indirectly from the consumption of electrical power. As the United States promulgates new policies for GHG emissions reporting and monitoring, it is becoming increasingly important to identify opportunities and solutions to GHG emissions management challenges, as well as develop a plan for compliance with emerging climate change policies. Because treatment plants are a quantifiable source of greenhouse gas emissions, there are potential permitting and reporting implications.

Reporting of GHG emissions and establishing a cap and trade program are two policy tools that have been under increased discussion over the past several years. Cap and trade is an environmental policy tool that places a mandatory cap on emissions, which is lowered over time in order to work toward emission reduction goals.

Ultimately, cap and trade programs may offer a cost-effective emissions reduction strategy. Cap-and-trade was the policy tool that stopped acid rain in the United States (U.S. EPA, 1995). The basic premise is to place a cap on emissions from different sources (emitters), and allow trades between sources in order to get environmental results at the lowest cost.

Most recently, cap-and-trade has been discussed in the United States in the context of reducing GHG emissions. A cap-and-trade program sets a clear limit on GHG emissions and minimizes the costs of achieving the target. Successful cap and trade programs reward innovation, efficiency and early action while providing strict environmental accountability without inhibiting economic growth. By creating a market – and a price – for emission reductions, cap and trade offers an environmentally effective and economically efficient response to lowering GHG emissions.

1.6 Previous GHG Emissions Studies

This evaluation addresses the relationship between GHG emissions and effluent discharge targets. Despite several studies that have quantified GHG emissions per treated wastewater (Keller and Hartley, 2002; Monteith et al., 2008; de Haas and Hartley, 2004), no study to our knowledge has considered the impact of stringent TN and TP limits on GHG emissions.

Quantifying the amount of GHG emissions with respect to flow for a wide range of nutrient discharge limits will provide insight into whether or not a point of diminishing returns is reached with respect to nutrient effluent targets and sustainability impacts in the form of GHG emissions. Other normalization strategies exist, such as load based.

A literature review of sustainability at WWTPs and how it relates to nutrient removal target objectives is discussed below with an emphasis on the following:

- Previous GHG emissions results from WWTPs
- The importance of establishing boundary conditions
- Impact of nutrient removal on WWTPs sustainability by way of GHG emissions

A summary of previous studies that have quantified GHG emissions from various WWTPs is provided in Table 1-1. The results range from 1,700 to 10,000 CO_2 equivalent metric tonnes per year (CO_2 eq mt/yr), for the referenced studies scaled to a 10 mgd influent flow. The wide range in values is based on the lack of defined inputs used for each study. Additionally, the normalization comparison is with respect to flow rather than load. A load-based approach accounts for variable wastewater strengths.

Table 1-1. Other GHG Emission Results with Values Scaled to 10 mgd Influent Flow.
(100,000 PE for 100 gal/d/cap)

Source	CO_2 eq mt /yr)	Treatment Configuration
Conventional Activated Sludge		
Tripathi, 2007 [i]	7,200	With Tertiary Filtration; no Digestion
Keller and Hartley, 2003	18,800	With Anaerobic Digestion
Cakir and Stenstrom, 2005	880	10 d SRT with Anaerobic Digestion
Cakir and Stenstrom, 2005	1,800	30 d SRT with Anaerobic Digestion
Cakir and Stenstrom, 2005	300	5 d SRT with Anaerobic Digestion
Anaerobic Treatment		
Cakir and Stenstrom, 2005	1,850	30 d SRT in Anaerobic Digestion
Keller and Hartley, 2003	7,800	Anaerobic Digestion
Nitrogen Removal		
HDR Engineering (Northern CA WWTP)	5,000	Tertiary Filtration 6 months/yr; Anaerobic Digestion and Energy Recovery
Biological Nutrient Removal		
de Haas and Hartley, 2004	5,300 – 7,400	With Anaerobic Digestion and Energy Recovery
Wallis-Lage et al., 2009	2,400	With Anaerobic Digestion and Energy Recovery
Tripathi, 2007 [ii]	8,000	With Anaerobic Digestion and Energy Recovery
Tripathi, 2007 [iii]	10,000	With Anaerobic Digestion and Energy Recovery
Hunt et al., 2009	2,250	Only considers activated sludge process (Johannesburg)
Hunt et al., 2009	2,250	Only considers activated sludge process (MLE)
Hunt et al., 2009	1,720	Only considers activated sludge process (Step BNR)
Hunt et al., 2009	2,270	Only considers activated sludge process (Step Feed)
Hunt et al., 2009	2,950	Only considers activated sludge process (Multi Sludge)

Notes:
i Ann Arbor WWTP
ii Laguna WWTP
iii Ypsilanti Community Utilities Authority

1.7 System Inputs

The extents of calculated GHG emissions at WWTPs are highly variable as shown in Table 1-1. The variability is partially due to the lack of consensus on system inputs. For example, the energy production to operate the WWTP is not always incorporated into the GHG emissions reported in all these studies. GHG emissions associated with manufacturing and distributing the chemicals used for treatment also is not necessarily included in GHG emissions calculations either. The existing non-uniform approach complicates the issue while attempting to compare studies.

The Climate Registry (2008) proposed a means to define emissions boundary conditions by placing them into three different "scopes" labeled 1, 2, or 3 emissions. The criteria for those three emissions groupings are outlined below:

- **Scope 1 – Direct on-site GHG emissions. These are generally described as anthropogenic GHG emissions (those related to man's activities):**
  - On-site production of electricity, heat or steam using fossil fuels
  - Physical or chemical processing
  - Transportation of materials, products, waste and employees
  - Fugitive emissions
  - Natural gas combustion

  For WWTPs, important Scope 1 emissions are presented in Table 1-2.

Table 1-2. Scope 1 - Direct GHG Emissions from Wastewater Treatment Processes.

WWTP Process	Scope 1 - GHG Emissions
Primary Treatment	None expected. Fugitive CH_4 (considered minimal)
Secondary Treatment	None expected from well managed aerobic/trickling filter process operating in carbonaceous BOD removal mode
Nutrient Removal	N_2O from nitrification-denitrification biological processes
Solids Handling	CH_4 from incomplete combustion of digester gas. Fugitive CH_4 (considered minimal)
Fugitive Emissions	Intentional/unintentional releases of GHG emissions from joints, seals, gaskets, etc. (e.g. methane leaks from digesters)
Heat/Power generation	Natural gas combustion or fossil fuel used for use in boilers, prime movers, HVAC units, and power generation

Carbon dioxide produced in the biological treatment processes and from cogeneration from biogas is considered biogenic and is not included as Scope 1 emissions. Biogenic carbon is part of the natural carbon balance and it will not add to atmospheric concentrations of carbon dioxide (IPCC, 2006).

- **Scope 2 – Indirect GHG emissions from imports (or purchases) of electricity, heat or steam.**

  For most WWTPs, this is dominated by purchased electricity.

- **Scope 3 – Other Indirect Supply-Chain GHG emissions. These GHG emissions sources are consequences of the activities of the reporting entity, but occur from sources owned or controlled by another entity:**
  - Transportation of products, materials and waste
  - Outsourced activities
  - Emissions from waste
  - Emissions from final product disposal

- Employee commuting
- Production of imported materials, such as chemicals

For WWTPs, the dominant Scope 3 focal areas are biosolids hauling/reuse, waste hauling, chemical manufacturing and hauling, and employee commute.

1.8 Future Regulations

Since the Clean Water Act became law in 1972, it has served as the foundation in surface water quality protection. In 1998, the U.S. EPA set out to set numeric nutrient criteria in all states. The U.S. EPA's goal at the time was to implement nutrient numeric standards in all states by 2002. Although currently unsuccessful at meeting this goal, several states have made notable progress since the U.S. EPA's inception of their goal. It is anticipated that the U.S. EPA's vision of a quantitative nutrient discharge standard will be put into action over the next several years. A detailed discussion on the regulatory outlook can be found in Chapter 3.0

Regulations and Driving Forces.

1.9 Impact of Nutrient Removal on GHG Emissions at WWTPs

This subsection will provide a background on the fundamentals of nutrient removal in WWTPs. The background is followed by a brief overview on the difficulty in meeting stringent nutrient removal discharge values per nutrient, such as TN <2 mg N/L, TP <0.02 mg P/L). Following the WWTPs nutrient removal background, greenhouse gas emissions at WWTPs are discussed and how GHG emissions and nutrients are related. For example, reverse osmosis (RO) or similar membrane technology is recommended in meeting the most stringent nutrient removal criteria in Level 5 (TN <2 mg N/L, TP <0.02 mg P/L).

1.9.1 Nutrient Removal

The removal of nitrogen from WWTPs is primarily achieved by (a) assimilation of nitrogen into biomass and (b) biochemical oxidation/reduction processes that convert organic nitrogen and ammonia to nitrogen gas through a two-step process. The two-step process is commonly referred to as nitrification and denitrification, whereby nitrification entails the oxidation of ammonia to nitrate by nitrifying organisms:

(1)

Nitrification is followed with denitrification by denitrifying bacteria that utilize organic carbon as an electron donor to reduce nitrate to nitrogen gas:

(2)

Many combined nitrification and denitrification configurations are employed in wastewater treatment plants to reduce total nitrogen (TN) effluent discharge levels to 4-8 mg N/L. A more detailed discussion on nitrification/denitrification, as well as the various secondary treatment configurations, can be found in the WEF Nutrient Removal Manual (2010). Oxidation and conversion of ammonia to nitrogen gas using nitrite is used to treat high strength wastewater using the anammox bacteria. This process is emerging as an attractive option for treating dewatering return streams.

The other nutrient of interest – phosphorus – can be removed from WWTPs biologically and/or by chemical/physical means. Biological phosphorus removal methods involve (a) assimilation of phosphorous (macro-nutrient) into cellular mass, and (b) enhanced phosphorus uptake by culturing phosphorus accumulating organisms (PAOs) in a sequential anaerobic-aerobic process.

The latter is done in tertiary processes, such as the Bardenpho, A2O, VIP, and other configurations. The accumulation of phosphorus by PAOs increases the concentration phosphorus in the wasted activated sludge. A detailed discussion on the mechanisms associated with biological phosphorus removal can be found in the WEF Nutrient Removal Manual (2010).

The elevated phosphorus levels are released from the biomass during anaerobic digestion, which promotes favorable conditions for struvite ($MgNH_4PO_4 \cdot 6H_2O$) formation (Doyle and Parsons, 2002). Struvite crystal formations can clog pipes, which in turn increase pumping time(s) and increases pumping cost, reduce plant capacity, and can be costly if pipe replacement is necessary.

Struvite control is heavily dependent on reducing ammonia, magnesium and phosphate concentrations coupled with pH control. An increase in pH reduces struvite solubility and the potential for precipitation. An alternative strategy to reducing nuisance struvite formation is to intentionally form struvite at a controlled location and subsequently recover struvite as a potential fertilizer.

In addition to biological P removal, chemical precipitation using a metal salt, such as alum or ferric, is used to reliably meet low effluent TP levels (TP <0.1 mg P/L). Chemical addition will precipitate inorganic phosphorus. The extent of phosphorus precipitation is largely governed by pH, alkalinity, and the metal dose to orthophosphate ratio.

The appropriate pH operating range is based on the coagulant used. Recent work has shown that the metal/phosphorus chemistry is closely tied to hydroxide formation and that covalent bonding to metal hydroxides forms the major mechanism for phosphorus removal. A more detailed discussion on the fundamentals of phosphorus chemical precipitation can be found in Smith et al. (2008).

1.9.1.1 Benefits of Longer Solids Retention Time

The activated sludge operator must increase the solids retention time (SRT) when transitioning from BOD to BNR mode as ammonia is oxidized by the slow growing nitrifying bacteria. The growth rate of nitrifying bacteria is the limiting factor. This increase in SRT has additional benefits beyond improving the discharge water quality, such as:

- Lower BOD/TSS discharged
- Higher removal of trace organic compounds (TrOCs) and heavy metals (Horz et al., 2004)
- Higher removal of heavy metals
- Water more suitable for reuse applications, in particular Level 5 (Chan et al., In press)
- Less algal blooms as there is less nutrient load discharged
- Secondary clarifier effluent better conditioned for filtration and disinfection (Leu et al., In press; Chan et al.; In press)

- Greater process stability from the anaerobic/anoxic zones serving as a selectors (Jenkins et al., 2004)

The increased removal of TOrCs is a function of increased surface area for sorption, elevated biomass concentration, and the enzymes that carry out nitrification (Horz et al., 2004). The enzyme responsible for the oxidation of ammonia to nitrite, ammonia monooxygenase, is commonly referred to as 'promiscuous' because it has the ability to assist in the biodegradation of a wide-range of compounds such as TOrCs (Vader et al., 2000).

The ability to filter secondary clarifier effluent is improved as the floc size/structure is a function of SRT. The number of fine particles leaving a secondary clarifier is inversely related to SRT. Sludge surfaces at high SRTs are less negatively charged and subsequently more hydrophobic (Liao et al., 2003), thereby creating less fine particles.

By releasing less fine particles, the secondary clarifier effluent is more filterable. Recent research by Chan et al. (In press) developed a novel indicator, mean particle size (MPS), that integrates particle size over three different particle diameters (0.5 to 50, 100, and 200 microns). Their analyses support the notion that MLSS floc sizes increase with SRT, and that the number of particles leaving the secondary clarifiers as supernatant decreased with increased SRT.

A corresponding study by Leu et al. (In press) further supports these claims as they found the average particle number in a secondary plant (<2 d SRT) versus a nitrification/denitrification plant (10-12 d SRT) was 29.2 x 10^5 cells/mL versus 1.3 x 10^5 cells/mL, respectively.

1.9.2 Role of Dissolved Organic Nitrogen on Permits

A portion of the nitrogen load that enters a municipality is recalcitrant (i.e. very slowly degradable; resistant to biological and chemical treatment) and commonly referred to as refractory dissolved organic nitrogen (rDON) (Stensel et al., 2008). A survey of over 30 municipal wastewater treatment plants showed that the effluent DON concentrations range from 0.5 to 2.5 mg N/L (Pagilla, 2007), with values as high as 2.8 mg N/L. Given that some permits require total nitrogen discharge on the order of 1 mg N/L, DON concentrations can govern a discharger's ability to meet permit.

The chemical compounds that contribute to rDON are still largely unknown. Parkin and McCarty (1981) used various size exclusion techniques to characterize rDON, but they only identified roughly 30% of the compounds. Recent work by Pehlivanoglu-Mantas and Sedlak (2008) and Khan et al. (2007) attempted to better characterize and understand rDON. Their findings are largely comparable to the Parkin and McCarty results.

Although the composition of rDON is largely unknown, the measured compounds comprised primarily of dissolved free and dissolved combined amino acids (DFAA/DCAA; roughly 10-20% of effluent DON) (Pehlivanoglu-Mantas and Sedlak, 2006). Dimethylamine (DMA), ethylenediaminetetraacetic acid (EDTA), and nitrilotriacetic acid (NTA) make up the remaining proportion of measured compounds. The remaining ~70% of DON has not been identified and is thought to be bound to other compounds, such as N-functional groups bound to humics (hydrophobic). Refer to the WERF compendium on rDON by Stensel et al. (2008) for more details.

1.9.3 Role of Phosphorus Species on Permits

Phosphorus is comprised of various species, with the two primary forms as either particulate or soluble. Within each species, phosphorus species can be characterized in terms of the analytical method used to measure them.

Reactive P (e.g., ortho-P) includes phosphorus species that is measured using a reactive dye and primarily indicate orthophosphates; acid hydrolysable P includes phosphorus that is released under mild acid conditions (e.g. polyP); and total phosphorus includes all phosphorus species detected following a strong acid digestion step (APHA, 2005). As with nitrogen, there is a species of P that is recalcitrant, referred to as refractory dissolved organic phosphorus (rDOP). The compounds that comprise refractory phosphorus are largely unknown.

Currently, phosphorus removal technologies at WWTPs primarily target ortho-P, those compounds that can be converted to ortho-P, or particulates elimination using chemical and/or biological methods. As the discharge permits require more phosphorus removal, the non-reactive P species become increasingly important. For example, the recalcitrant dissolved organic phosphorus (rDOP) species plays a critical role since it is not removed biologically or chemically.

Full-scale performance data from various facilities indicate that biological phosphorus removal at WWTPs can achieve an effluent discharge from 0.1-0.3 mg P/L (Neethling et al., 2005; Gu et al., 2008; Drury et al., 2005; Liu and Gu, 2010). Chemical phosphorus removal by metal salts is commonly used to improve phosphorus removal and achieve low (below 0.5 mg P/L) effluent limits. According to the database and statistics performed by Bott and Parker (2011), a combination of biological and chemical removal can reliably meet a maximum monthly level of 0.1 mg P/L on a 95th percentile basis. To achieve lower levels such as Level 5 (<0.02 mg P/L) is dependent on the species of non-reactive soluble P, chemical dose, pH, and the ability to remove the precipitated solids. Based on these uncertainties, it is difficult to predict a range of values achievable by chemical precipitation/filtration.

1.9.4 Greenhouse Gas Emissions at Wastewater Treatment Plants

Some gaseous byproducts produced during municipal wastewater treatment are known GHG emissions contributors. The byproducts of interest have evolved from only including methane gas (CH_4) (IPCC, 2001) to the incorporation of CO_2, CH_4 (IPCC, 2006), and nitrous oxide gas (N_2O) (adaptation of Ahn et al., 2010b).

CO_2 created during the oxidation of cBOD does not constitute a GHG emissions contributor as the CO_2 is part of the natural carbon cycle and referred to as "biogenic carbon." The three areas that comprise the majority of GHG emissions from a municipality are CO_2 emissions from energy demand (e.g., pumping, mixing, and aeration), fugitive CH_4 in the collection system and anaerobic digestion, and N_2O emissions during biological treatment.

Outside of the energy demand associated with pumping, mixing and aeration, the corresponding subsections provide background information on other key areas of GHG emissions at WWTPs.

1.9.4.1 Energy Demand at Wastewater Treatment Plants

Energy is the largest contributor to GHG emissions at WWTPs (Keller and Hartley, 2007; Tripathi, 2007). As will be presented in Chapter 4.0, energy represents over 80% of the total

(Scope 1, 2, and 3) GHG emissions at a WWTP. At most plants, electrical energy dominates. However, many plants also use significant quantities of natural gas.

Electrical energy is a Scope 2 emission and the GHG emissions associated with electricity is from the CO_2, N_2O, and CH_4 produced by combustion of fossil fuels at the electric utility. Consumption of natural gas for process and building heat as well as power generation is a Scope 1 emission.

1.9.4.2 Methane Production in Wastewater Processes

Methane formation at WWTPs is largely ignored in GHG emissions calculations as suggested in the IPCC methane fugitive methane emissions empirically based multipliers (IPCC, 2006). Anaerobic conditions that promote methane formation within a WWTP may exist in the collection system, primary clarifiers, anaerobic selectors, improperly aerated aeration basins, anaerobic digesters (intentional), digested sludge dewatering (gas or in filtrate/centrate), and in the effluent stream (through BOD degraded anaerobically in receiving waters).

Despite these different locations for potential methane formation in a WWTP, it is thought to yield a minimal contribution within the boundaries of a WWTP. This report assumes the IPCC (2006) multipliers for all levels.

1.9.4.3 Nitrous Oxide Production in Wastewater Processes

The WERF Research Project U4R07a by Chandran et al. (2010) provides detailed information on N_2O emissions and the mechanisms associated with their formation. The measurement of N_2O gases has gained recent attention as a GHG emissions because it is about 300 times more potent than carbon dioxide and its greenhouse gas impact and has been detected with improved analytical monitoring (e.g., online analyzers). Besides its potency, N_2O has the innate ability to directly deplete the ozone layer (Ravishankara et al., 2009).

The change from secondary treatment to biological nutrient removal (BNR) could increase N_2O and nitric oxide (NO) gas emissions. Although N_2O and NO gases are both produced, emphasis is placed on N_2O in this report as the emission levels are two to three orders of magnitude greater than IPCC (2006) based on measurements performed by Chandran et al. (2010). N_2O reporting strategies typically rely on either a multiplier against flow (IPCC, 2006; U.S. EPA, 2009) or a multiplier against reactor influent TKN load (Chandran et al., 2010; Ahn et al., 2010b). The inherent flaw in using a multiplier against flow is that nitrogen loads are highly variable. For example, return side stream flows from centrate/filtrate following anaerobic digestion has an order of magnitude greater nitrogen concentration than the main plant flow. Rather than use a flow based multiplier, basing N_2O emissions on TKN load addresses the nitrogen concentration issue.

Table 1-3. N_2O Emissions from Various Full-Scale Studies (Ahn et al., 2010b).*

Process	Proportion of Influent TKN as N_2O (%)	Reference	Sample Collection
Activated Sludge (11,000 PE)	0.035	Czepiel et al., 1995	Grab and On-line
Nitritation-anammox	2.3	Kampschreur et al., 2008b	Grab
Separate nitrification stage (620,000 PE)	4.0	Kampschreur et al., 2008b	Grab
Activated Sludge (1,000 PE)	0.01-0.08	Kimochi et al., 1998	On-line
Activated Sludge (60,000 PE)	0.02	Sommer et al., 1998	Grab
Activated Sludge (60,000 PE)	0.001	Sümer et al., 1995	Grab
25 Different Activated Sludge Processes	0-15	Wicht and Beier, 1995	Grab
U.S. nationwide survey of 12 Different BNR and Non-BNR Activated Sludge Processes	0.01-1.8	Ahn et al., 2010b	On-line

*An average N_2O emissions for each grouping, non-BNR and BNR, were used in the analysis. The data range for the U.S. nationwide survey was used to illustrate the variability.

The study by Ahn et al. (2010b) includes data from 12 different activated sludge facilities in the United States. Their findings suggest that reactor configuration (e.g., MLE, oxidation ditch, step feed, etc.) plays a role in the concentration of N exiting the treatment plant as N_2O with values ranging from 0.03 (separate stage BNR) to 1.8 (step feed non-BNR) percent of the influent TKN load. Given the wide range of N_2O emission values, this study used the average N_2O emissions for non-BNR and BNR plants from the Ahn et al. (2010b) U.S. nationwide survey. The data range per data grouping (non-BNR and BNR) is presented as error bars to highlight the variability. The orders of magnitude difference in proportion TKN emitted as N_2O is attributed to several parameters:

- Sampling approach (grab versus on-line as shown in Table 1-3)
- Reactor configuration and operation levels (e.g., dissolved oxygen and ammonia) (Park et al., 2000; Tallec et al., 2008; Yu et al., 2010).
- Sampling time of day as N_2O emissions vary diurnally (Ahn et al., 2010a)
- Sample location within basin (Ahn et al., 2010a)

The IPCC (2006) and U.S. EPA (2009) methods assume that N_2O emissions only occur during denitrification in the anoxic zone. However, others have shown that nitrification can also lead to N_2O emissions. Figure 1-2 (Kampschreur at al., 2008) shows potential biological transformations pathways of nitrogen and how N_2O potentially forms during both nitrification and denitrification. The known mechanisms that can lead to N_2O emissions by AOB include the following:

- Oxygen limitation (Bock, 1995; Hooper et al., 1997; Schmidt and Bock, 1997; Schmidt and Bock, 1998)
- Dynamic conditions, especially from oxygen or ammonia concentrations (Kampschreur et al., 2008a; Kampschreur et al., 2008b; Kester et al., 1997)
- Recovery from anoxic conditions with higher emissions in aerobic zones (Yu et al., 2010)

Despite advancements in monitoring analyses and a further understanding on the mechanisms associated with N_2O emissions at the gene level, a unified emission factor and/or algorithm for calculating N_2O emissions for wastewater treatment processes is not available at this time. Rather than assume a particular BNR configuration for N_2O emissions and the corresponding emission factor from Ahn et al. (2010b), this study assumed the average for non-BNR (three plants in total) and BNR (eight plants in total) plants coupled with data range bars per sample set.

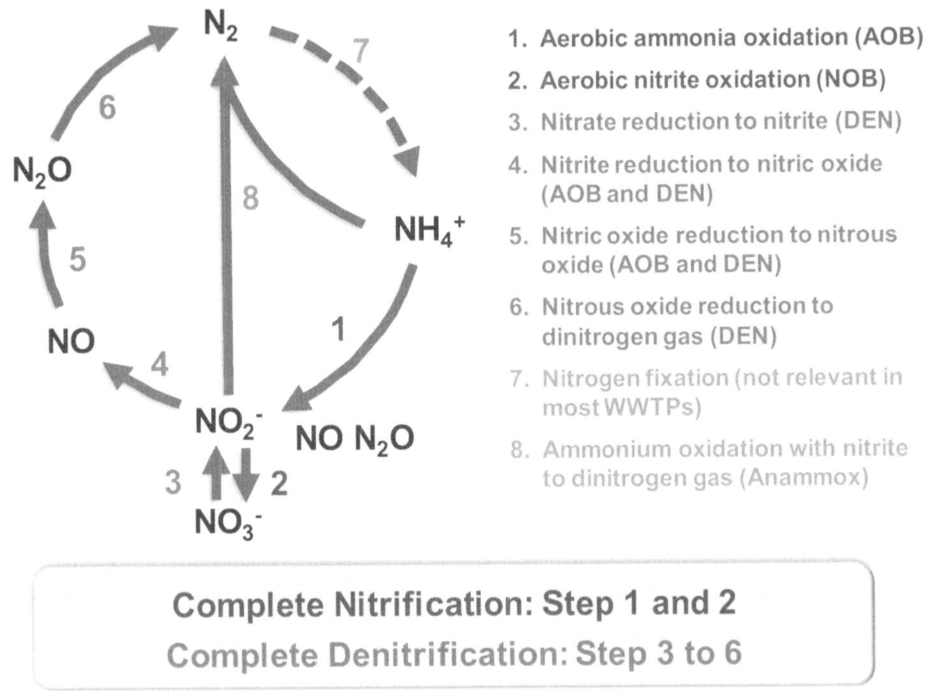

Figure 1-2. Biological Nitrogen Conversions.
N pathway adapted from Kampschreur et al., 2009.

1.9.4.4 External Carbon Sources

The ability to achieve low total nitrogen levels (TN < 8mg N/L) typically requires an external carbon source for denitrification. Historically, methanol has been the most commonly utilized carbon source for tertiary denitrification processes both for economic reasons and relatively low solids yields. During the past several years, there has been an increased focus on using alternative carbon sources for denitrification. The interest in alternative carbon sources is due to safety concerns associated with methanol, increases in methanol costs and lapses in its availability. There has been significant research into denitrification kinetics, biomass yields and operating issues associated with these carbon sources. Commonly used external carbon sources are as follows (Gu et al., 2010):

- Pure chemicals (e.g., methanol, ethanol, acetate, sugar, and butanol)
- Commercially available or manufactures (e.g., Unicarb, Micro C™, etc)
- Raw industrial/agricultural byproducts (e.g., corn syrup, molasses, brewery waste, etc.)
- Sludge fermentation products
- Electron donor gases (e.g., hydrogen gas, methane, etc.)

Research has expanded to understanding the denitrifying bacteria that perform denitrification and how they are impacted by different external carbon sources. Baytshtok et al.

(2008; 2009) found that *Methyloversatilis* spp. and *Hypomicrobium* spp. were the predominant bacteria in sequencing batch reactors fed methanol. Upon transitioning from methanol to ethanol as the external carbon source, *Hypomicrobium* spp. decreased in numbers, whereas *Methyloversatilis* spp. remained relatively unchanged as they have the ability to assimilate either methanol or ethanol. However, based on the overall decrease in concentrations for both bacterial species, the denitrification rate was reduced.

Lu and Chandran (2010) compared the predominant bacteria for suspended versus attached growth systems associated with denitrification for a sequencing batch integrated fixed-film activated sludge (SB-IFAS) process fed glycerol.The predominant bacteria associated with denitrification varied for the suspended growth versus the attached growth bacteria. In particular, the suspended bacteria were comprised of *Comamonas* spp. and *Diaphorobacter* spp., whereas the attached growth denitrifiers were predominantly comprised of *Comamonas* spp., *Bradyrhizobium* spp., and *Tessaracoccus* spp.

This report assumes methanol is used as the external carbon source for all levels of treatment requiring an external carbon source (Levels 3, 4, and 5). Although methanol is a natural fuel, it is fossil based when produced on the scale used a WWTPs. Because it is fossil based, it is considered a GHG emission in WWTP audits and factored into the calculations. The chemical production of methanol yields 0.67 lb CO_2 per lb methanol (Willis et al., 2009). This GHG emissions value does not include the hauling of chemicals from the manufacturing WWTP to its end use location.

1.9.4.5 Chemicals

Wastewater treatment requires a variety of chemicals for flocculation, coagulation, disinfection, solids thickening and dewatering, and a carbon source for growth. In particular, WWTPs with low level nitrogen and phosphorus permits require an external carbon source and/or metal salt. Although the contribution of chemicals to GHG emissions is excluded in Scope 1 and 2 emissions, they are included in Scope 3 emissions.

GHG emissions associated with chemical usage (production and distribution) at WWTPs are often overlooked. It is critical that the amount of GHG emissions associated with each individual chemical during production is incorporated into the evaluation. Several chemicals are mined in a few locations globally and the mining and transportation of the chemicals contribute to global GHG emissions. For example, the closest ferric mine to the United States is in Jamaica. For treatment plants located on the west coast, Jamaica is several thousand miles away and requires hauling. Additionally, the distance travelled, fuel type and truck fuel efficiency all play a role in quantifying their respective GHG emissions. This report uses the chemical production hauling values from Tripathi (2007) and the IPCC (2007).

The GHG emissions impact of an external carbon source such as methanol must account for production, distribution, and CO_2 emissions during its utilization. A recent study by Kang et al. (2009) highlights the role of an external carbon source while performing GHG emissions analyses on WWTPs. They found that BNR plants that use methanol increase GHG emissions at the plant between 25 to 100% increase.

1.9.4.6 Brine Reject Management

For the most stringent permit levels (e.g., <2 mg N/L, <0.02 mg P/L), a technology that can remove rDON and rDOP is required (e.g., RO). Despite on-going research investigating rDON and rDOP removal with advanced oxidation processes (AOPs) that breakdown the compounds to bioavailable forms or membranes less tight than RO, the industry standard technology RO was assumed. Less tight membranes than RO, such as nanofiltration (NF), have been found to remove about 80% of the more difficult to remove nitrogen species that were previously thought to require RO (Xu et al., 2010). Although promising, NF membranes for rDON removal is considered leading edge and thus excluded from the analysis. The brine reject of RO requires an energy intensive brine management solution, such as deep well injection, that more than doubles the overall GHG emissions for a WWTP.

The RO process concentrates contaminants into a smaller volume reject stream. Disposing of the RO reject stream can be a problem because of the potentially large volume of water involved and the concentration of contaminants contained in the brine. The primary treatment/handling options for RO reject are:

- Surface water discharge
- Ocean discharge
- Haul and discharge to coastal location for ocean discharge
- Sewer discharge
- Deep well injection
- Evaporate in a pond
- Solar pond concentrator
- Advanced thermal evaporation

A brief description for each reject stream management alternative is provided below. For a more detailed breakdown on sizing and their capital construction costs, please refer to Mickley et al. (2006).

Surface Water Discharge

For areas that offer large tracts of affordable acreage (e.g., rural areas), surface water discharge can be a cost-effective method for reject stream disposal. Implementation typically requires storage (above-ground or within a lined pond). The storage equalizes discharge flow, provides solids settling, and/or cools the waste discharge. Surface water discharge does not treat the reject stream per se because it provides an area for the water to be driven off to the atmosphere and reduce the reject stream volume. Once concentrated, the dried product still requires handling and the approach should only be considered as an intermediate step in the handling process.

Ocean Discharge

The discharge of reject stream to the ocean is limited to coastal locations. If a reject stream, commonly referred to as a "brine line" is readily available, this might offer the most effective reject stream handling option. If no reject stream discharge line is in place, the construction and permitting timeline can be timely.

Sewer Discharge

Discharging to a sewer is a common brine disposal method in urbanized areas and can be equal or less than the cost of surface water discharge. This approach is limited to small industrial flows or smaller drinking water systems. Because this report focuses on wastewater treatment, sewer disposal is not a viable option.

Deep Well Injection

Deep well injection (DWI) of reject brine is widely used in the oil industry. Deep well injection can provide a safe means of permanent brine disposal because the brine is sequestered from potential drinking water aquifers. Some desalination facilities, as well as oil and gas companies, have rejected the DWI method because these wells are difficult to permit and maintain. A Class 1 or Class 5 nonhazardous well permit (as defined by the U.S. EPA) would be required for brine purge disposal. Very high pressures (and energy) also may be required for deep well injection.

Evaporative Pond

Evaporation ponds offer a reliable, "low-tech" method to concentrate the brine and potentially reduce hauling and disposal costs. Although effective, evaporation ponds are considered to be land intensive. However, evaporation ponds can be used in tandem with other technologies, such as thermal concentrators, to reduce land requirements.

Driving factors for evaporation ponds are area demands, flows, flow composition and meteorological conditions. It is estimated that the high strength waste stream (> 20% total dissolved solids (TDS)) will evaporate at a 30% slower rate than fresh water and consequently requires more acreage on a flow per unit area basis. Although evaporative ponds are reliable, local climate conditions, urban area land needs, and relatively high land costs can reduce their viability.

A key issue for brine management strategies is how to deal with the final concentrated salt product. At some point, the evaporative pond reaches capacity and requires the salt be managed. Typically, the salt is crystallized, packaged in plastic containers and disposed of by hauling to a landfill.

Salinity Gradient Solar Pond

The Salinity Gradient Solar Pond (SGSP) provides an alternative, less land intensive approach to evaporation ponds. While evaporative ponds attempt to drive moisture away by solar evaporation, SGSPs are designed to harness the solar power within the pond to power a multi-staged evaporative unit, such as a multi-effect distillation (MED) unit. In addition to a concentrate stream, a condensation stream (i.e., distilled water) is collected with the MED unit that can be used for other on-site reuse applications. Similar to the evaporative pond, dealing with the final concentrated salt product is problematic.

Advanced Thermal Evaporation

Advanced thermal evaporation requires a two-step process consisting of a brine concentrator followed by crystallizer. A brine concentrator followed by crystallizer is the most compact approach for concentrating the brine purge stream. The crystallized product can be hauled to a landfill for ultimate disposal.

Summary

There are many techniques which can be used to manage brine associated with RO reject. The appropriate alternative is primarily governed by geographic and local constraints. A comparison of the relative brine management costs are provided in Table 1-4.

Table 1-4. Brine Disposal Method Relative Cost Comparison.

Disposal Method	Relative Capital Cost	Relative O&M Cost	Comments
Surface Water Discharge	Lowest	Lowest	Both capital and O&M costs heavily dependent on the distance from brine generation point to discharge. Not an option for nutrient removal.
Sewer Discharge	Low	Low	Both capital and O&M costs heavily dependent on the brine generation point to discharge distance. Higher cost than surface water discharge due to ongoing sewer connection charge. Not an option for wastewater treatment.
Deep Well Injection	Medium	Medium	Technically sophisticated discharge and monitoring wells required. O&M cost highly variable based on injection pumping energy.
Evaporation Ponds	Low – High	Low	Capital cost highly dependent on the amount and cost of land.
Salinity Gradient Solar Ponds	Low – High	Lowest	Same as evaporation ponds plus added cost of heat exchanger and pumps. Lower O&M cost due to electricity production.
Advanced Thermal Evaporation	High	Highest	Extremely small footprint, but the energy from H_2O removal is by far the most energy intensive unless waste heat is used.

Of the listed options, the capital costs associated with DWI is comparable nationwide as it requires drilling and pumping. In contrast, an evaporation pond cost is dependent on the acreage costs. Although we assumed DWI for the analysis, it is important to recognize that the GHG emissions vary widely for the seven brine management options listed above.

1.10 Example of GHG Emissions Inventory Performed on WWTP

An example of Scope 1 and 2 emission results for a municipality is provided in this subsection as a means to show the steps associated with performing a GHG emissions audit on a municipality. The example used is for a 15 mgd WWTP located in California. The GHG emissions and carbon footprint boundary for this analysis is defined as operations within the property bounds.

1.10.1 Accounting and GHG Emissions Boundary Conditions for a 15 mgd WWTP

Table 1-5 below outlines the accounting for GHG emissions at a WWTP in California.

Table 1-5. Summary of GHG Emissions Accounting at the WWTP.

GHG Emission Type	GHG Emission Calculated
Scope 1 - Direct – Stationary	CO_2 eq emissions from testing/operating on-site generators Emissions from natural gas combustion
Scope 1 - Direct – Mobile	CO_2 eq emissions from plant operations and maintenance vehicles
Scope 1 - Direct – Process	N_2O from nitrification and NDN processes; CH_4 from anaerobic digester flare inefficiencies; Cogeneration produced by digester gas
Scope 2 – Indirect	CO_2 eq from purchased electricity

Calculations of these GHG emissions are summarized below.

1.10.2 Emission Calculations

The emissions were calculated using the Local Government Operations Protocol (Ver. 1.1) (California Air Resources Board et al., 2008) and the Climate Registry General Reporting Protocol (Ver. 1.1; 2008). For the purposes of the analyzed inventories, default CO_2 eq multiplication factors of 23 and 296 where used for CH_4 and N_2O, respectively, per the above protocols.

Scope 1 - Direct Stationary – Direct stationary emissions at the WWTP are limited to natural gas and the emergency standby generators used for back-up power. Natural gas is used for building heating in the administration/laboratory building, maintenance building and boiler/heat exchanger.

Scope 1 - Direct Mobile – Direct mobile emissions at the WWTP were calculated for on-site vehicles associated with routine operations and maintenance work at the WWTP. An estimate for daily miles traveled both in the base and future years was used to develop GHG emissions.

Scope 1 - Direct Process – The liquid treatment process GHG emissions include nitrification and denitrification (NDN) processes in the activated sludge system. The anaerobic digester gas produced is either used by the cogeneration process or flared, depending on the inventory scenario. These types of digester gas flares offer some inefficiency in combustion of the gas which is considered in the GHG emissions calculations.

Scope 2 - Indirect – Purchased Electricity – Indirect emissions related to the WWTP operations include the purchase of electricity, which is typically the most significant GHG emissions from wastewater treatment facilities. To determine baseline power use, historical energy consumption data was evaluated for the WWTP. Recent historical energy use reflects current operations and supports development of the baseline (Inventory 1) indirect GHG emissions.

1.10.2.1 Methodology

Inventories were developed for the various Scope 1 emissions (digester gas flaring, mobile sources, stationary combustion, and nitrification) and Scope 2 emissions (electrical consumption). CO_2, CH_4, and N_2O emissions were determined using the above protocols. The equivalent CO_2 emissions were subsequently calculated using standard protocol conversions (see Basis of Calculations below). The total equivalent CO_2 emissions were then compiled using contributions from each source and summarized at the end of this document. Scope 1 and Scope 2 emissions were analyzed for these inventories.

1.10.2.2 Basis of Calculation

The inventory calculations were based on the following:

- Fuel, electric and digester gas consumption increases by a factor of 1.67 from the 9 mgd to 15 mgd levels
- The digester gas electrical cogeneration facility will be operational during inventory numbers 2 and 3
- The digester gas electrical cogeneration facility would not be operational during inventories 4, 5 and 6
- The CO_2 equivalent multiplier is 23 for CH_4 and 296 for N_2O

1.10.3 Base Year and Future Years Used for GHG Emissions

A GHG emissions inventory was developed for different alternatives as shown in Table 1-6. The alternatives consider conditions in 2008, followed by future flows and loads in year 2014 for nitrogen removal with and without cogeneration. The conversion from secondary treatment (Inventory 1) to NDN (Inventory 2) increases the GHG emissions by 10% (4,690 increased to 5,170 CO_2 eq mt/yr). If cogeneration is used, the increase in GHG emissions related to a NDN conversion (Inventory 3) is off-set and actually lower than secondary treatment (Inventory 1; 4,690 reduced to 4,550 CO_2 eq mt/yr). Based on the GHG emissions analysis, cogeneration was incorporated into the WWTPs design. This GHG emissions evaluation highlights how a particular technology can govern the design selection process.

Table 1-6. Base Year and Future GHG Emissions Calculation Scenarios for a 15 mgd WWTP.

Inventory	Time Period (annual average flow)	Description of Treatment Scenario of Influent Wastewater to the WWTP	GHG Results Scope 1 and 2 (CO_2 eq mt/yr)
1	Base Yr 2008 (9 mgd)	Flow treated at the WWTP; cBOD mode; No cogeneration; Existing treatment process	4,690
2	Future Yr 2014 (9 mgd) NDN Upgrades Complete	Flow treated at the WWTP; No cogeneration; NDN added to treatment process	5,170
3	Future Yr 2014 (9 mgd) NDN Upgrades Complete	Flow treated at the WWTP; Cogeneration added; NDN added to treatment process	4,550

CHAPTER 2.0

REGULATIONS AND DRIVING FORCES

2.1 Introduction

Nutrient removal treatment can reduce point source discharges of nitrogen and phosphorus. Substantial investments are required to build and operate advanced wastewater treatment facilities. Nitrogen and phosphorus loadings to watersheds impact water quality by stimulating the growth of algae, which may result in depletion of dissolved oxygen, shifts in pH, degradation of habitat, impairment of drinking water sources, and in some cases harmful algal blooms (U.S. EPA, 2010).

According to the U.S. EPA, nearly every state has nutrient-related pollution with impacts in over 80 estuaries/bays, and thousands of rivers, streams and lakes. In particular, the U.S. EPA cites the Gulf of Mexico and the Chesapeake Bay as examples of significant water quality impacts from 35 states that contribute to nutrient loadings in the Gulf and Bay.

Nutrient loadings from both point and non-point sources contribute to water quality impairments in the nation's waterways. Point source discharges from wastewater treatment plants can be a significant source of nitrogen and phosphorus in watersheds. Non-point sources also contribute substantial amounts of nutrients from land use activities such as agriculture, forestry, and urban/suburban development.

In some watersheds, non-point source nutrient loadings outweigh point sources to a degree that advanced treatment for nutrient removal, and even complete elimination of point sources, would have limited benefit to water quality. Nevertheless, point source national pollution discharge elimination system (NPDES) permitted dischargers are the most directly regulated sources subject to nutrient control requirements resulting from numeric nutrient standards, total maximum daily loads (TMDLs), and water quality-based permit limits.

2.2 U.S. EPA Numeric Nutrient Standards

The U.S. EPA initiated an effort in 1998 to adopt numeric nutrient standards in all states by 2002. This goal has not been accomplished. On May 25, 2007, U.S. EPA Assistant Administrator Ben Grumbles issued a memorandum titled "Nutrient Pollution and Numeric Water Quality Standards" to state and tribal water program directors. The memo underscores the U.S. EPA 's commitment to accelerate the development of numeric nutrient water quality standards. Some states and territories have made notable progress since then, but overall results have been inconsistent. U.S. EPA recently released water quality standards for the state of Florida (U.S. EPA, 2010).

Grumbles noted in the 2007 memo that high nitrogen and phosphorus loadings result in harmful algal blooms, reduced spawning grounds and nursery habitats, fish kills, oxygen-starved hypoxic or "dead" zones, public health concerns related to impaired drinking water sources, and increased exposure to toxic microbes, such as *Cyanobacteria sp*. Descriptions of four waterbodies of particular importance are included in the memo:

- Chesapeake Bay. This watershed has an existing hypoxia problem and adds 150,000 new people per year to the watershed.
- The Gulf of Mexico. There are 31 states that contribute to the watershed. This watershed has a prevalent and well-documented hypoxic "dead" zone presumed to be caused by nitrogen pollution. However, in the 2001 Hypoxia Action Plan, the U.S. EPA Science Advisory Board reports that phosphorus plays a much greater role in the hypoxia problem than previously thought.
- Long Island Sound. Dissolved oxygen (DO) is below standards in one-third to one-half of the Sound. Nitrogen loadings have been capped at 1990 loads, and a water quality trading program has been implemented in Connecticut for point sources with a market-based approach.
- Puget Sound. The Puget Sound Action Plan calls for a better understanding of nutrient and bacteria loadings from septic systems.

2.3 Stakeholder Input to the Nutrient Impact Discussion – NRDC and Others

The national discussion of nutrient impacts on water quality is evolving and is highlighted in high visibility waterbodies such as the Chesapeake Bay, Long Island Sound, the Gulf of Mexico and Puget Sound, as indicated above. Several organizations have entered the discussion and some have called for treatment technology standards for nitrogen and phosphorus.

In November 2007, the Natural Resources Defense Council (NRDC) filed a Petition for Rulemaking with the U.S. EPA to redefine secondary treatment and limit nutrient pollution from wastewater treatment facilities. Ten other regional and national environmental groups, including the Sierra Club and American Rivers, joined NRDC in the petition. NRDC argues that nitrogen and phosphorus effluent limitations should be a part of the base technology definition of secondary treatment. NRDC contended that the U.S. EPA must protect the public by establishing nutrient limits, specifically that the U.S. EPA unreasonably delayed publishing information on secondary treatment to remove nutrients. The NRDC also notes that nutrient control is properly included within "Secondary Treatment" and cites the following effluent nutrient levels as attainable:

- Effluent TP at 1.0 mg P/L and TN at 8.0 mg N/L are consistently attainable using current biological based technologies
- Effluent TP at 0.5 mg P/L and TN at 5 mg N/L are consistently attainable using current biological based technologies that might require complementary chemicals (e.g., external carbon source for denitrification)
- Effluent TP at 0.3 mg P/L and TN at 3.0 mg N/L is attainable with existing technology using only improved biological and chemical treatment processes

The NRDC considers the U.S. EPA's approach to site-specific nutrient standards as unreasonable and believes the U.S. EPA cannot rely on a water-quality based approach to control nutrient pollution. NRDC argues that nutrient pollution is widespread and justifies a generally-applicable standards approach. NRDC calls for the U.S. EPA to specify the degree of nitrogen and phosphorus reduction attainable through secondary treatment.

The National Association of Clean Water Agencies (NACWA) provided comments on the NRDC Petition for Rulemaking on Secondary Treatment in a letter to the U.S. EPA dated Feb. 29, 2008 (NACWA 2008). NACWA expressed concern that the NRDC petition calling for nutrient limits as part of the secondary treatment process is not technically or financially practical, and that the approach is not the most effective or environmentally sensitive way to reduce nutrient pollution. NACWA criticize the proposed "one size fits all" approach to a water quality problem that is site-specific and best to address site-specific measures.

NACWA believes site specific water quality efforts will be more effective than technology based nutrient removal limits and cites five areas of concern with the NRDC petition:

1. Legal basis for incorporating nutrient removal into secondary treatment.
2. Failure of the petition to address the contribution to nutrient loadings from non-point sources.
3. Potentially high costs for treatment plants to meet a national nutrient limit and whether such expenditures are cost-effective.
4. Increased negative environmental impacts of mandating a national nutrient removal limit.
5. Inappropriateness of national limits for local and regional water quality issues.

NACWA research notes that tertiary treatment for nutrient removal goes well beyond the original Congressional intent in the Clean Water Act for secondary treatment and that U.S. EPA has denied previous requests to include nutrient removal as part of secondary treatment standards. NACWA officials have said that the petition fails to acknowledge the impact of non-point sources of nutrients on water quality. NACWA members have found that the costs for nutrient removal are substantial and that major expenditures are associated with expanding facilities because of the need for more plant site space to accomplish retrofits.

2.4 The Ecoregion Concept and Nutrient Criteria

The concept of ecological regions, or ecoregions, is the grouping of areas of similar climate, hydrology, geology, physiography, soils, land use, vegetation and wildlife. The ecoregion criteria concept was included in the National Strategy for the Development of Regional Nutrient Criteria from the beginning, citing the work of James Omernik of the U.S. EPA's laboratory in Corvallis, Oregon. There are four levels of ecoregions where each level has a different level of detail in grouping. For example, Level I has the broadest resolution as it divides North America into 15 different regions. In contrast, Level IV has the finest resolution with over 100 different regions within North America. An example of a Level III ecoregions map of continental United States is shown in Figure 2-1 (104 different regions).

The U.S. EPA has established in stream criteria for total phosphorus and total nitrogen for Level I ecoregions for rivers and streams, lakes and reservoirs, and wetlands. The nutrient criteria presented by the U.S. EPA for each ecoregion are generally based on the 25th percentile value of all data from the respective ecoregion. The 25th percentile value corresponds to the concentration at which 25% of the measured values are below and 75% of the measured values are above (U.S. EPA, 2000). A summary of the rivers and streams criteria are shown in Table 2-1.

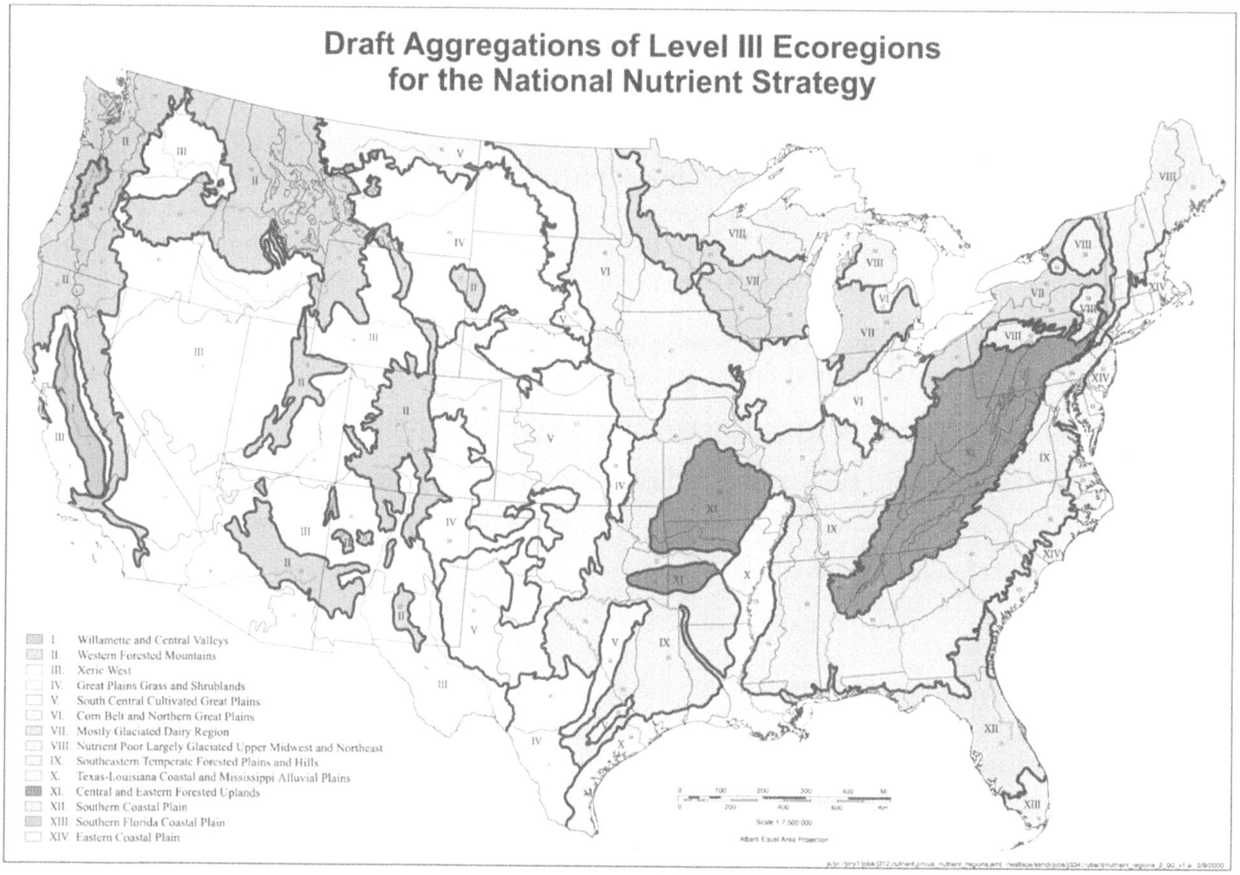

Figure 2-1. U.S. EPA Level III Ecoregions.

Table 2-1 summarizes Ecoregion Criteria for Rivers and Streams with and without dilution. The values in Table 2-1 are presented in two columns per constituent with the first supplying the in-stream nutrient level and the second supplying the value discharged from the WWTP for a 20:1 dilution. The 20:1 dilution credit was used because a less dilute discharge is considered a wastewater dominated stream. The corresponding required treatment level for the 20:1 dilution is also provided in Table 2-1.

Table 2-1. Summary of Ecoregion Criteria for Rivers and Streams With and Without Dilution.

Ecoregion	TN (mg N/L) IN-STREAM	TN (mg N/L) DISCHARGE (20:1 Dilution)	TP (mg P/L) IN-STREAM	TP (mg P/L) DISCHARGE (20:1 Dilution)	Treatment Level (20:1 Dilution)
I: Willamette and Central Valley	0.66	13.2	0.055	1.1	Level 2
II: Western Forested Mountains	0.12	2.4	0.010	0.2	Level 5
III: Xeric West	0.38	7.6	0.022	0.44	Level 3
IV: Great Plains Grass and Shrublands	0.56	11.2	0.023	0.46	Level 3
V: South Central Cultivated Great Plains	0.88	17.6	0.067	1.34	Level 2
VI: Corn Belt and Northern Great Plains	2.18	43.6	0.076	1.52	Level 2
VII: Mostly Glaciated Dairy Region	0.54	10.8	0.033	0.66	Level 3
VIII: Nutrient-Poor, Largely Glaciated Upper Midwest and Northeast	0.38	7.6	0.010	0.2	Level 3
IX: Southeastern Temperature Forested Plains and Hills	0.69	13.8	0.037	0.74	Level 3
X: Texas-Louisiana Coastal and Mississippi Alluvial Plains	0.57	11.4	0.060	1.2	Level 2
XI: The Central and Eastern Forested Uplands	0.31	6.2	0.010	0.2	Level 4
XII: Southeastern Coastal Plain	0.90	18	0.040	0.8	Level 3
XIII: Southern Florida Coastal Plain	1.14	22.8	0.015	0.3	Level 3
XIV: Eastern Coastal Plain	0.71	14.2	0.031	0.62	Level 3

2.5 Effluent Technology Limits Do Not Guarantee Water Quality

Increased levels of treatment for nutrient removals do not guarantee receiving water quality improvements. Water quality is impacted by many factors with complex relationships in the aquatic environment, including the magnitude of pollutant loading compared to the receiving waters, limiting nutrients that control aquatic growth, and receiving water physical characteristics.

Much focus has been placed on point sources for control of nutrient discharges, where point sources are only a part of the nutrient loading picture. The EPA now recognizes that most watersheds are impaired by a combination of point and nonpoint sources. In fact, EPA data indicates that point sources may be contributing only a small fraction of water quality impairment as shown in Figures 2-2, 2-3, and 2-4.

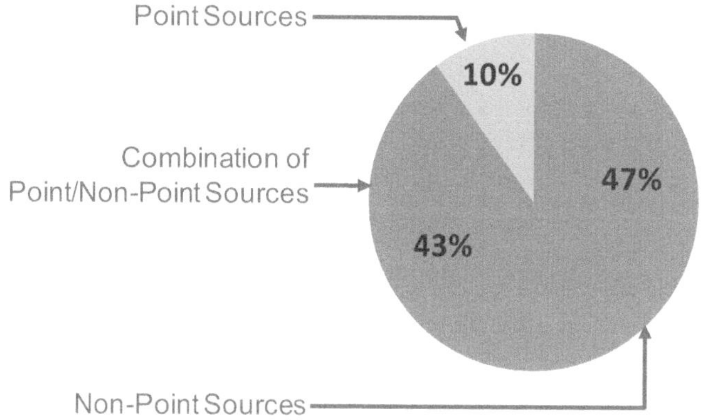

Figure 2-2. Sources of Nationwide Water Quality Impairment (EPA, 2001).

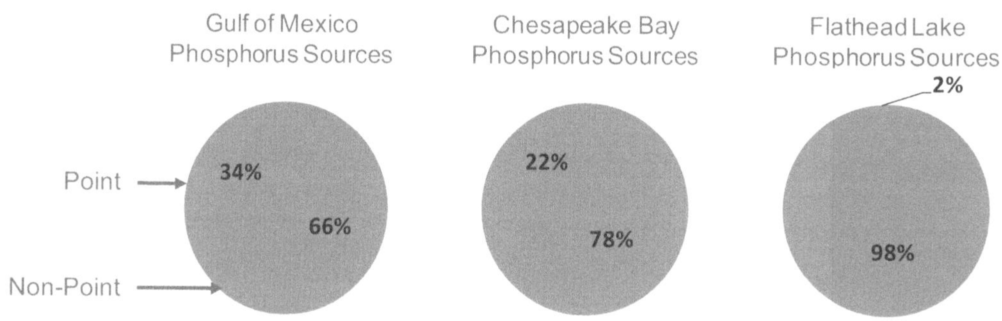

Figure 2-3. Phosphorus Loading Summaries for Gulf of Mexico, Chesapeake Bay and Flathead Lake.
Sources: Gulf of Mexico Hypoxia 2008 Action Plan, Chesapeake Bay Program Action Plan, Montana Department of Environmental Quality, Draft Nutrient Management Plan and Total Maximum Daily Load for Flathead Lake, Montana.

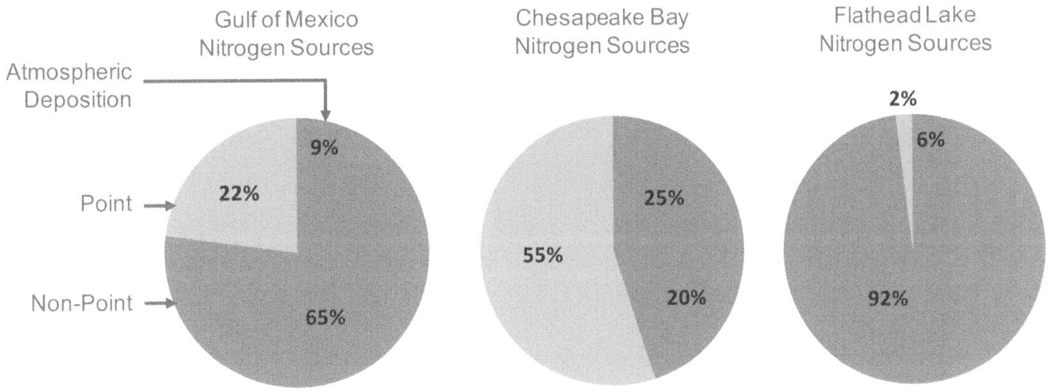

Figure 2-4. Nitrogen Loading Summaries for Gulf of Mexico, Chesapeake Bay, and Flathead Lake.
Sources: Gulf of Mexico Hypoxia 2008 Action Plan, Chesapeake Bay Program Action Plan, Montana Department of Environmental Quality, Draft Nutrient Management Plan and Total Maximum Daily Load for Flathead Lake, Montana.

Reducing point sources alone may have minimal impact of receiving water quality. Costs and sustainability impacts dramatically increase for treatment facilities with little added value to the environment. A comprehensive approach to point and non-point nutrient discharges will yield the most practical solution to improving water quality in our watersheds.

CHAPTER 3.0

FIVE TIERED LEVELS OF TREATMENT, DESIGN CRITERIA, GHG EMISSIONS CONVERSION ASSUMPTIONS, AND SYSTEM INPUTS

The objective of this study is to determine the sustainability impacts as wastewater treatment plants implement treatment technologies to meet increasingly stringent nutrient limits. Five levels of treatment are considered for reducing nitrogen and phosphorus from secondary to levels approach in stream nutrient criteria. This section presents the treatment levels and the process flow schemes that could be implemented on a Greenfield plant. It also presents the criteria used to determine the sustainability impacts.

3.1 Five Tiers Concept

Five different target objectives were developed as shown in Table 3-1. A baseline case was included, referred to as Level 1, which is a carbonaceous removal WWTP with primary, secondary, and disinfection on the liquid stream. As for Levels 2 to 5, all facilities have tertiary treatment in the form of biological nutrient removal, plus additional tertiary physical/chemical treatment technologies, depending on the effluent objective. Full biological nitrification is assumed for Levels 2 to 5. The nutrient removal values selected are based on the future wide range of values anticipated by stakeholders, such as the U.S. EPA, NRDC, grassroots organizations, etc.

Table 3-1. Treatment Levels Objectives.

Level	Biochemical Oxygen Demand (mg/L)	Total Suspended Solids (mg/L)	Total Nitrogen as Nitrogen (mg N/L)	Total Phosphorus as Phosphorus (mg P/L)
1	30	30	-	-
2	<30	<30	8	1
3	<30	<30	4-8	0.1-0.3
4	<30	<30	3	0.1
5	<30	<30	<2	<0.02

Treatment unit processes associated with each level are shown in Table 3-2. Schematics and background of each individual treatment level are provided in subsections.

Table 3-2. Unit Processes per Treatment Level.

Level	Liquid Treatment	Solids Treatment	Comments
1	Primary Clarifier Activated Sludge Disinfection Dechlorination	Gravity Belt Thickener Anaerobic Digestion with Cogen Centrifugation	Conventional Activated Sludge for BOD/TSS removal
2	Primary Clarifier Activated Sludge Alum (optional) Disinfection Dechlorination	Gravity Belt Thickener Anaerobic Digestion with Cogen Centrifugation	Nitrification/Denitrification and Biological Phosphorus Removal
3	Primary Clarifier Activated Sludge Methanol (optional) Alum (filtration) Filtration Disinfection Dechlorination	Gravity Belt Thickener Anaerobic Digestion with Cogen Centrifugation	Nitrification/Denitrification and Biological Phosphorus Removal and Filtration
4	Primary Clarifier Activated Sludge Methanol (optional) Alum/Polymer (Enhanced Settling) Enhanced Settling Filtration Disinfection Dechlorination	Fermentation Gravity Belt Thickener Anaerobic Digestion with Cogen Centrifugation	Nitrification/Denitrification and Biological Phosphorus Removal, High Rate Clarification and Denitrification Filtration
5	Primary Clarifier Activated Sludge Methanol (optional) Alum/Polymer (Enhanced Settling) Enhanced Settling Filtration Microfiltration Reverse Osmosis Disinfection Dechlorination	Gravity Belt Thickener Anaerobic Digestion with Cogen Centrifugation	Nitrification/Denitrification and Biological Phosphorus Removal, High Rate Clarification, Denitrification Filtration, and MF/RO on about Half the Flow

The treatment options reflect conventional technology in 2010. Clearly, newer technologies will emerge as the industry is pushed to treatment Level 3 or beyond. Some newer technologies are already used and emerging. A paradigm shift could completely change the outlook on wastewater treatment. For example, nutrient resource recovery and source separation are emerging as more sustainable and cost-effective options.

3.1.1 Level 1 (30 mg/L BOD; 30 mg/L TSS; no nutrient requirements)

Level 1 serves as a base case with current conventional secondary treatment with traditional cBOD and TSS removal facilities. A schematic of treatment Level 1 is shown in Figure 3-1.

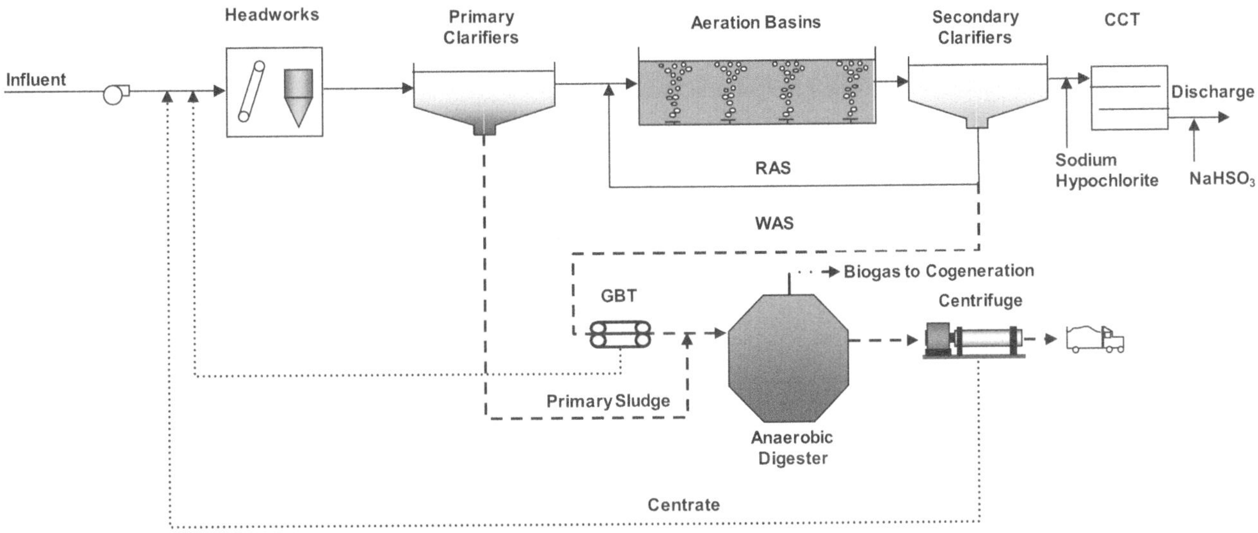

Figure 3-1. Level 1 Plant Schematic.
(30 mg/L BOD; 30 mg/L TSS; no nutrient requirements)

Besides primary and secondary treatment, disinfection prior to discharge, as well as solids treatment, is provided. Solids treatment entails thickening by gravity belt thickener (GBT), anaerobic digestion, and dewatering by centrifuge. The biogas produced during anaerobic digestion will be used as an energy source for cogeneration using an engine with a jacket and exhaust heat recovery.

3.1.2 Level 2 (8 mg N/L; 1 mg P/L)

The Level 2 WWTP represents the least stringent of four nutrient removal facilities. A schematic of treatment Level 2 is shown in Figure 3-2. The secondary treatment WWTP consists of a conventional type biological nutrient removal process, whereby the first zone is anaerobic, followed by anoxic and aerobic zones. In this evaluation a three-stage Phoredox (A2O) style process is used for representative biological treatment.

The mixed liquor return (MLR) carries nitrified water from the aerobic phase to the anoxic phase for denitrification. Alum addition is optional prior to the secondary clarifiers as an additional barrier since the water goes from secondary treatment directly to disinfection, followed by discharge. As for solids treatment, the unit processes are comparable to the base case with thickening by GBT, anaerobic digestion, and dewatering by centrifuge. The biogas produced during anaerobic digestion will be used as an energy source for cogeneration using an engine with a jacket and exhaust heat recovery.

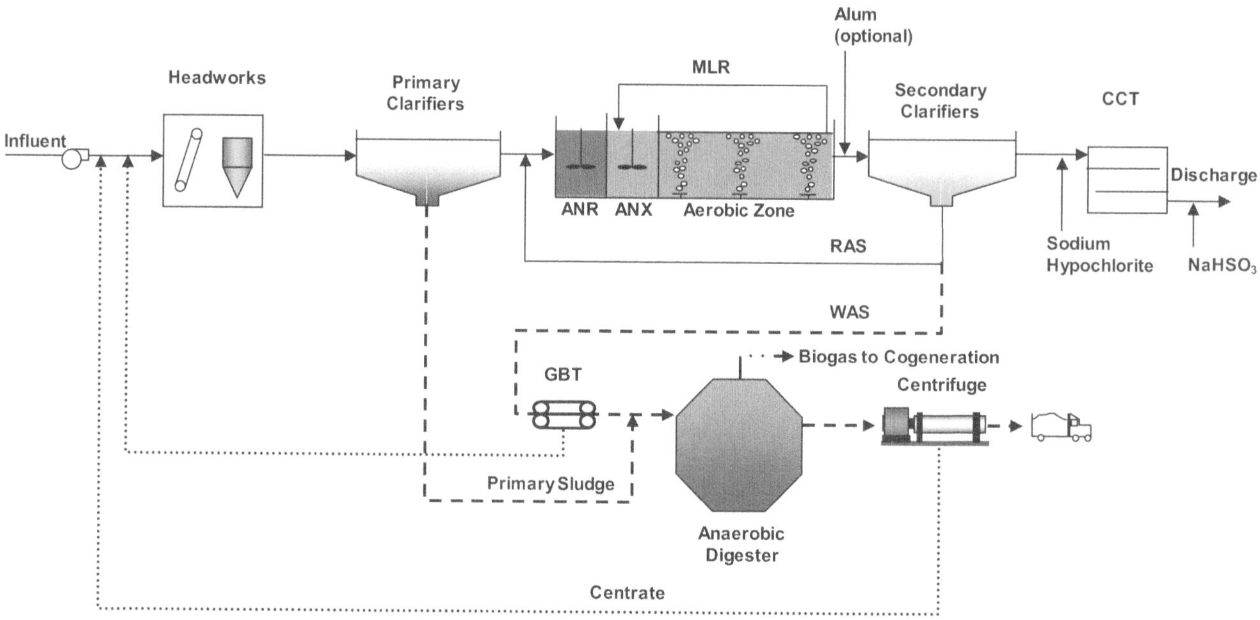

Figure 3-2. Level 2 Plant Schematic.
(8 mg N/L; 1 mg P/L)

3.1.3 Level 3 (4-8 mg N/L; 0.1-0.3 mg P/L)

The nutrient discharge levels are more stringent than Level 2 and require additional removal mechanisms. A schematic of treatment Level 3 is shown in Figure 3-3. The secondary process transitions from the A2O process in Level 2 to a 5-stage Bardenpho in Level 3. The secondary process also is equipped with an optional external carbon source (e.g., methanol). The optional metal salt (e.g., alum) is relocated from the secondary clarifiers to the tertiary filtration process. The optional metal salt is not included in the GHG emissions calculations.

As with Level 1 and 2, the solids treatment contains a GBT, anaerobic digestion and dewatering by way of centrifuges. Also, the biogas produced during anaerobic digestion will be used as an energy source for cogeneration using an engine with a jacket and exhaust heat recovery.

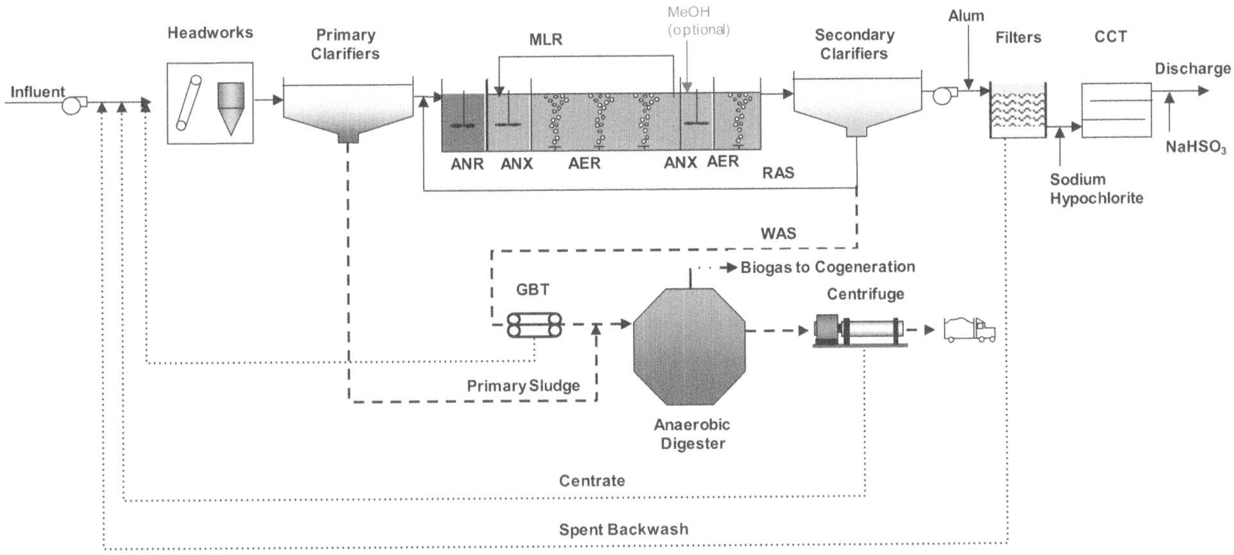

Figure 3-3. Level 3 Plant Schematic.
(4-8 mg N/L; 0.1-0.3 mg P/L)

3.1.4 Level 4 (3 mg N/L; 0.1 mg P/L)

The WWTP in Level 4 requires more chemicals than Level 3 and larger biological reactors. A schematic of treatment Level 4 is shown in Figure 3-4.

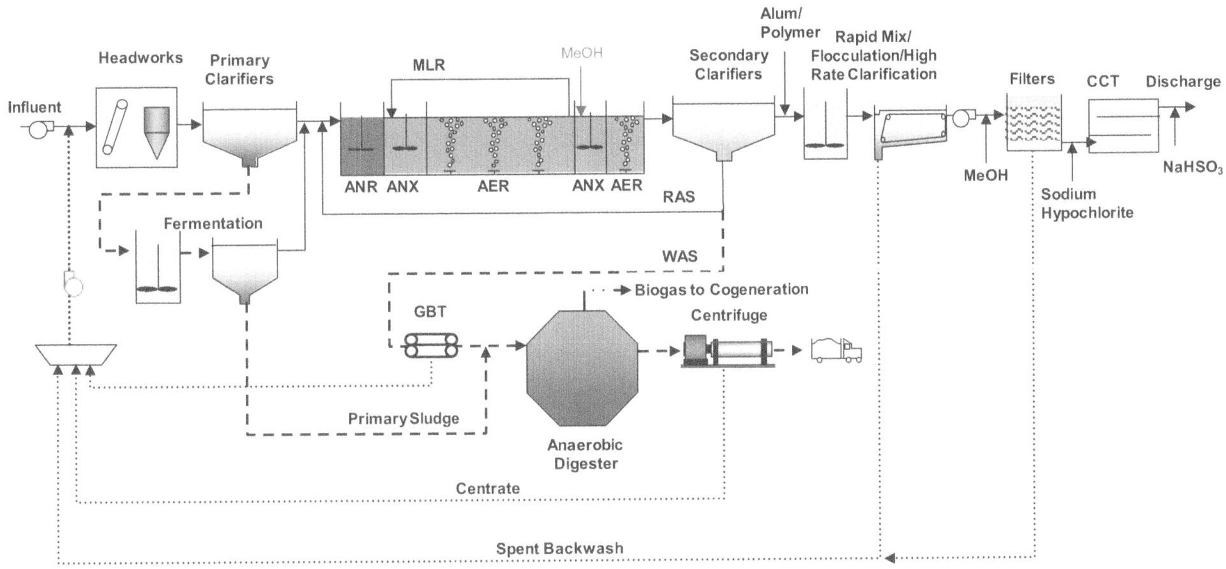

Figure 3-4. Level 4 Plant Schematic.
(3 mg N/L; 0.1 mg P/L)

The current influent wastewater characteristics may not consistently supply sufficient soluble BOD (sBOD) to fuel N and P removal to meet Level 4 objectives. To overcome this potential limitation, primary solids are subjected to fermentation. The fermented primary solids produce volatile fatty acids (i.e., sBOD) that are returned to the liquid stream for the secondary process. The fermentation step requires fermentation equipment as well as the plumbing and pumps associated with routing the flow. The secondary process is similar to Level 3 except that methanol addition is required. Following secondary treatment, a high rate clarification (HRC) process is located upstream of the filters to promote the precipitation and removal of P. To ensure the removal of nitrate, the filter serves as a denitrification filter.

The solids treatment in Level 4 differs from the previous levels. The primary solids are fermented to generate sBOD for the secondary process. Additionally, the return side streams (RSS) are equalized and slowly bleed back into the liquid stream. The equalization of RSS is critical given the fact that RSS constitutes roughly 20% of the overall nitrogen load on the secondary process. Centrate treatment is an option, such as single stage Anammox, would reduce the energy/chemical requirements for treating that 20% plant N load. As with all the previous levels, the biogas produced during anaerobic digestion will be used as an energy source for cogeneration using an engine with a jacket and exhaust heat recovery.

3.1.5 Level 5 (2 mg N/L; <0.02 mg P/L)

The most stringent permit of the five-tiered approach requires advanced tertiary treatment technologies, such as a membrane filtration. A schematic of treatment Level 5 is shown in Figure 3-5. The removal strategy mimics Level 3 through the secondary clarifiers. Following the secondaries, a 50-50 split occurs where half the flow is sent through high rate clarification and denitrification filters and the other half is sent through side-stream RO. Side-stream RO is implemented to meet the stringent <2 mg N/L and <0.02 mg P/L. The first membrane, microfiltration (MF), serves as pretreatment for the second membrane step, reverse osmosis (RO). The RO is required to meet the permit due to reduce the rDON/rDOP to meet the target effluent levels. A less tight membrane, such as NF might achieve the necessary rDON/rDOP removal to meet the discharge permit. However, that approach is considered an emerging process and thus not considering in the analysis.

The solids treatment is comparable to Level 3. As with the previous levels, the cogeneration will use an engine with a jacket and exhaust heat recovery.

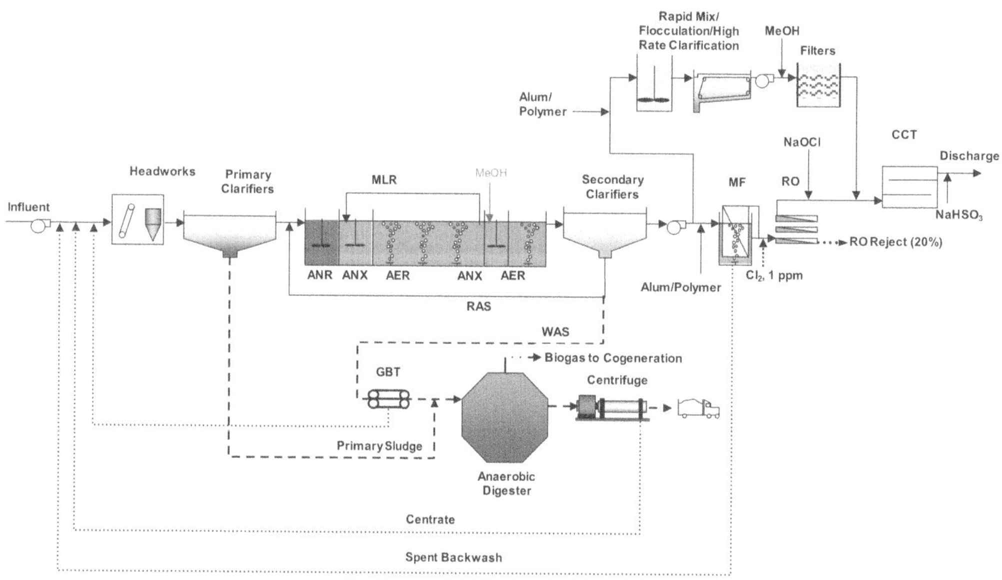

Figure 3-5. Level 5 Plant Schematic.
(2 mg N/L; <0.02 mg P/L)

3.2 Design Criteria

Design of the steady state mass balance is based on a 10 mgd annual average (AA) flow condition (100,000 population equivalents) and a normal peaking factor of 3 (WEF, 2009). Variability in flows and loads at any WWTP is a challenge that governs process sizing. In this analysis, AA condition and a summer temperature condition were used to compare the different levels of treatment. The use of AA condition is not standard engineering design practice to design a full-scale 10 mgd plant. The liquid and solid streams design criteria are listed in Tables 3-3 and 3-4, respectively.

Table 3-3. Liquids Stream Design Criteria – Based on Average Flow and Load.

Process	Design Criteria
Influent Flow	10 mgd = 100,000 population equivalents (PE) at 100 gal//d/capita
Temperature	20°C
Primary Clarifiers	1,000 gpd/sf; 2 hr HRT
Activated Sludge	-
Level 1	Aerobic SRT = 2 d, MLSS = 1,250 mg/L, DO = 2 mg/L, Alpha = 0.30
Level 2-5	Aerobic SRT = 10 d, MLSS = 2,500 mg/L, DO = 2 mg/L, Alpha = 0.45, MLR = 3Q
Secondary Clarifiers	600 gpd/sf, 24 lb/d/sf
High Rate Clarifier	200 sf/mgd
Filtration	3 gpm/sf on average flow
Denitrification Filter	1.5 gpm/sf on average annual flow
Microfiltration	50 gfd
Reverse Osmosis	40 gfd
Chlorine Contact Tank	30 – 60 min HRT

Table 3-4. Solids Stream Design Criteria.

Process	Design Criteria
Gravity Belt Thickener	275 gpm/m; 1,000 lb/hr/m
Anaerobic Digester	18 d (for Annual Average); 0.15 lb VSS/cf/d
Dewatering Centrifuges	10,000 lb/d (for Annual Average); 100 gpm (for Annual Average)
Fermentation	4 d (for Annual Average)

3.3 Greenhouse Gas Calculation Assumptions

The steady state mass balance results were used to calculate GHG emissions. The assumptions used to convert between energy demand, chemical demand and production, as well as biologically-mediated gases (i.e., CH_4 and N_2O) and GHG emissions is provided in Table. The assumptions are based on U.S. EPA (2007) values for energy production, an adaptation of the database provided in Ahn et al. (2010b) for N_2O emissions contribution, IPCC (2006) for fugitive CH_4 emissions, and various resources for chemical production and hauling from production to the WWTP. Additionally, the biogas produced during anaerobic digestion that is used as a fuel source is converted to energy with MOP8 (2009) recommended waste-to-energy values.

The IPCC (2006) and U.S. EPA (2009) do not adequately quantify N_2O emission factors. Both approaches assume that N_2O emissions only occur during denitrification in the anoxic zone. Nitrification can also lead to N_2O emissions. The mechanisms associated with N_2O emissions were described in Section 1.9.4.3. Despite advancements in monitoring, analyses, and a better understanding of the mechanisms associated with N_2O emissions at the gene level, a unified emission factor and/or algorithm for calculating N_2O emissions for wastewater treatment processes is not available at this time.

The N_2O emission factors are based on the average values per grouping (non-BNR and BNR) for the U.S. nationwide survey compiled by Ahn et al. (2010b) that investigated non-BNR and BNR. The average value for non-BNR (three WWTPs in total) and BNR (three WWTPs in total) were used. Furthermore, to account for the wide variability in the nationwide survey, the data range is provided per sample set. The Ahn et al. (2010b) values used in this study for non-BNR (Level 1) were 32 g N/PE/yr (data range from 5-97 g N/PE/yr) and for BNR (Levels 2 to 5) 30 g N/PE/yr (data range from 0.3-140 N/PE/yr).

Emission factors for sodium bisulfite are not readily available in the literature. Rather than leaving the chemical blank, we followed the approach implemented by Biswas (2009). Biswas accounted for sodium bisulfite by adding the emission factors for the two primary ingredients (i.e., caustic soda and sulfur dioxide) to determine the corresponding emission factor. It is important to recognize that the true emission factor for sodium bisulfite will be higher since the weighted emission factors used do not reflect the energy associated with combining the two.

Table 3-5. Greenhouse Gas Production Assumptions.

Parameters	Units	Value	Source
N_2O to CO_2 Conversion	lb CO_2/lb N_2O	296	IPCC, 2006
CH_4 to CO_2 Conversion	lb CO_2/lb CH_4	23	IPCC, 2006
Energy Production			
CO_2	lb CO_2/MWh	1,329	USEPA (2007)
N_2O	lb N_2O/GWh	20.6	USEPA (2007)
CH_4	lb CO_2/GWh	27.3	USEPA (2007)
Sum Energy Production	lb CO_2/MWh	1336	USEPA (2007)
GHGs per BTU Natural Gas			
CO_2	lb CO_2/MMBTU Natural Gas	52.9	CA Climate Action Registry Reporting Tool
N_2O	lb N_2O/MMBTU Natural Gas	0.0001	CA Climate Action Registry Reporting Tool
CH_4	lb CO_2/MMBTU Natural Gas	0.0059	CA Climate Action Registry Reporting Tool
Sum Natural Gas		53.1	CA Climate Action Registry Reporting Tool
Non-BNR N_2O Emissions	g N_2O/PE/yr	32	Ahn et al. (2010b)
BNR N_2O Emissions	g N_2O/PE/yr	30	Ahn et al. (2010b)
Biogas Purity	% Methane	65	WEF, 2009
Biogas to Energy	BTU/cf CH4	550	WEF, 2009
Digester Gas to Electrical Energy Transfer Efficiency	%	32	HDR Data
Chemical Production			
Alum	lb CO_2/lb Alum	0.28	SimaPro 6.0 - BUWAL250, Eco-indicator 95
Alkalinity	lb CO_2/lb Alkalinity	0.35	SimaPro 6.0 - BUWAL250, Eco-indicator 95
Polymer	lb CO_2/lb Polymer	1.18	Owen (1982)
Methanol	lb CO_2/lb Methanol	0.67	Willis et al., (2009)
Soda Ash	lb CO_2/lb Soda Ash	0.42	Assumes Trona Ore used, IPCC, 2006
Sodium Bisulfite	lb CO_2/lb Sodium Bisulfite	0.25	Biswas (2009)
Sodium Hypochlorite	lb CO_2/lb Sodium Hypochlorite	1.07	Owen (1982)
Building Energy Efficiency	kBTU/sf/yr	60	Calif. Commercial End-Use Survey (2006)
Hauling Distance		-	
Local	miles	100	-
Regional	miles	750	-
National	miles	2000	-
Hauling Emissions			
Fuel Efficiency	miles per gallon	8	
CO_2	kg CO_2/gal diesel	10.2	CA Climate Action Registry Reporting Tool
N_2O	kg N_2O/gal diesel	0.0001	CA Climate Action Registry Reporting Tool
CH_4	kg CH_4/gal diesel	0.003	CA Climate Action Registry Reporting Tool
Sum Hauling Fuel	kg CO_2/gal diesel	10.2	CA Climate Action Registry Reporting Tool

3.4 Incremental Increase in GHG Emissions per Additional Pound Nutrient Removed

To better understand the sustainability impacts with increased treatment, the incremental increase in GHG emissions was calculated per additional pound of nutrient removed. The incremental impact was calculated by taking the delta of the treatment level of interest and the previous level divided by the delta in nutrient discharge of the treatment level of interest and the previous level as follows:

$$\frac{\text{Incremental GHG Increase}}{\text{Additional lb Nutrient Removed}} = \frac{GHG_{\text{Level } i} - GHG_{\text{Level } i\text{-}1}}{\text{Nutrient Discharge Load}_{\text{Level } i} - \text{Nutrient Discharge Load}_{\text{Level } i\text{-}1}} \quad (3)$$

Each level of treatment was evaluated to determine which processes are required for the removal of that particular nutrient and the corresponding GHG emissions calculated. For example, the methanol demand and additional aeration requirements associated with Levels 2 to 5 is solely for nitrogen removal and thus excluded in the incremental GHG emissions increase for P removal. In contrast, the filtration and required chemicals (alum/polymer) is for P removal and thus excluded from the incremental GHG emissions for N removal. For instances where a process is used to remove both nitrogen/phosphorus, such as MF/RO, the process was fully weighted for each nutrient.

3.5 Economic Variables Assumptions

The economic variables used to calculate total present worth include the following:

- Life-span = 20 years
- Discount rate = 5.0%
- Escalation rates for capital, energy, and non-energy inflation rates = 3.5%

3.6 System Inputs

GHG emission outputs are highly variable largely in part due to a lack of consensus on system inputs. The sustainability impacts of WWTPs must also include chemical consumption, production of nitrous oxides and methane (both potent greenhouse gases), biosolids haul/reuse, and emissions (particularly of CH_4) created in the wastewater collection system. Some of these items are difficult to measure, and only limited data exists for many contributions including fugitive methane escaping the plant, nitrous oxide production in the activated sludge basins, and methane entering the plant with raw sewage.

System inputs are summarized in Table 3-6 with a corresponding plot in Figure 4-6. Note that the evaluation considers GHG emissions outside the boundary of a municipality. For example, both the biosolids and chemical hauling are included in the GHG emissions calculations. Additionally, GHG emissions associated with each chemical used in treatment is incorporated in the calculations. Although the GHG emissions calculation includes emissions generated during chemical production, the GHG emissions produced during the exploration of chemicals are not included in the calculation.

The miscellaneous values include the following:

- Odor control – 100 hp
- Non-potable water – assumed 5% of influent plant flow at 75 psi (70% efficiency)

- High pressure air system – 25 hp
- Miscellaneous pumping/mixing/sludge storage/transfer pumps – 75 hp
- Administrative buildings energy demand – assume 14,000 sf for 10 mgd flow at 60 kBTU/sf/yr (CEUS, 2006)
- Miscellaneous equipment – 75 hp

Table 3-6. System Inputs and Assumptions for this Investigation.

GHG Scope	GHG Contributor	Addressed in this Investigation	Reference/Value/Comment
1	Carbon dioxide and other GHG emissions produced from cogeneration	Not considered	Biogenic carbon (carbon of biological origin under the natural carbon cycle); Contributors outside cogeneration are small
1	Carbon Dioxide produced in wastewater treatment	Not considered	Biogenic carbon (carbon of biological origin under the natural carbon cycle)
1	Fugitive methane from anaerobic processes	Considered in calculation	Relatively small and considered for all conditions investigated. Default value from Climate Registry only considers septic tank and wetlands/lagoons.
1	Nitrous oxide discharged to the air and effluent	Considered in calculation	Considerable research now being done on Nitrous oxide emissions (e.g., Yu et al., 2010; Ahn et al., 2010a; 2010b; Kampschreur et al., 2009). Average values from U.S. nationwide survey from both non-BNR and BNR plants used in the analysis (Ahn et al., 2010b)
1	Natural gas for backup digester heating, building heating, other fuels used in plant	Considered in calculation	
1	Reuse of biogas on site in boilers and cogeneration	Not considered	Biogenic carbon (carbon of biological origin under the natural carbon cycle)
2	Miscellaneous onsite treatment include odor control, non-potable water, high pressure air system, miscellaneous pumping/mixing/sludge storage/transfer pumps, HVAC, administrative buildings energy, and miscellaneous equipment.	Considered in calculation	
2	Purchased electricity	Considered in calculation	
3	Manufacturing and hauling of chemicals	Considered in calculation	Coagulants, flocculants, methanol and other chemicals related to nutrient removal
3	Emissions associated with methanol used as external carbon source	Considered in calculation	
3	Biosolids hauling	Considered in calculation	Fuel for trucking. Offsite GHG emissions for reuse or disposal not included. Hauling included according to Climate Change Protocol.
3	Brine management from reverse osmosis	Considered in calculation	Treatment alternatives are wide that ranges from sewer discharge to advanced thermal evaporation. We used deep well injection.
3	Methane in raw sewage	Not considered	May be a significant contributor to GHG emissions in WWTPs but is considered essentially zero using IPCC (2006) for all treatment levels.

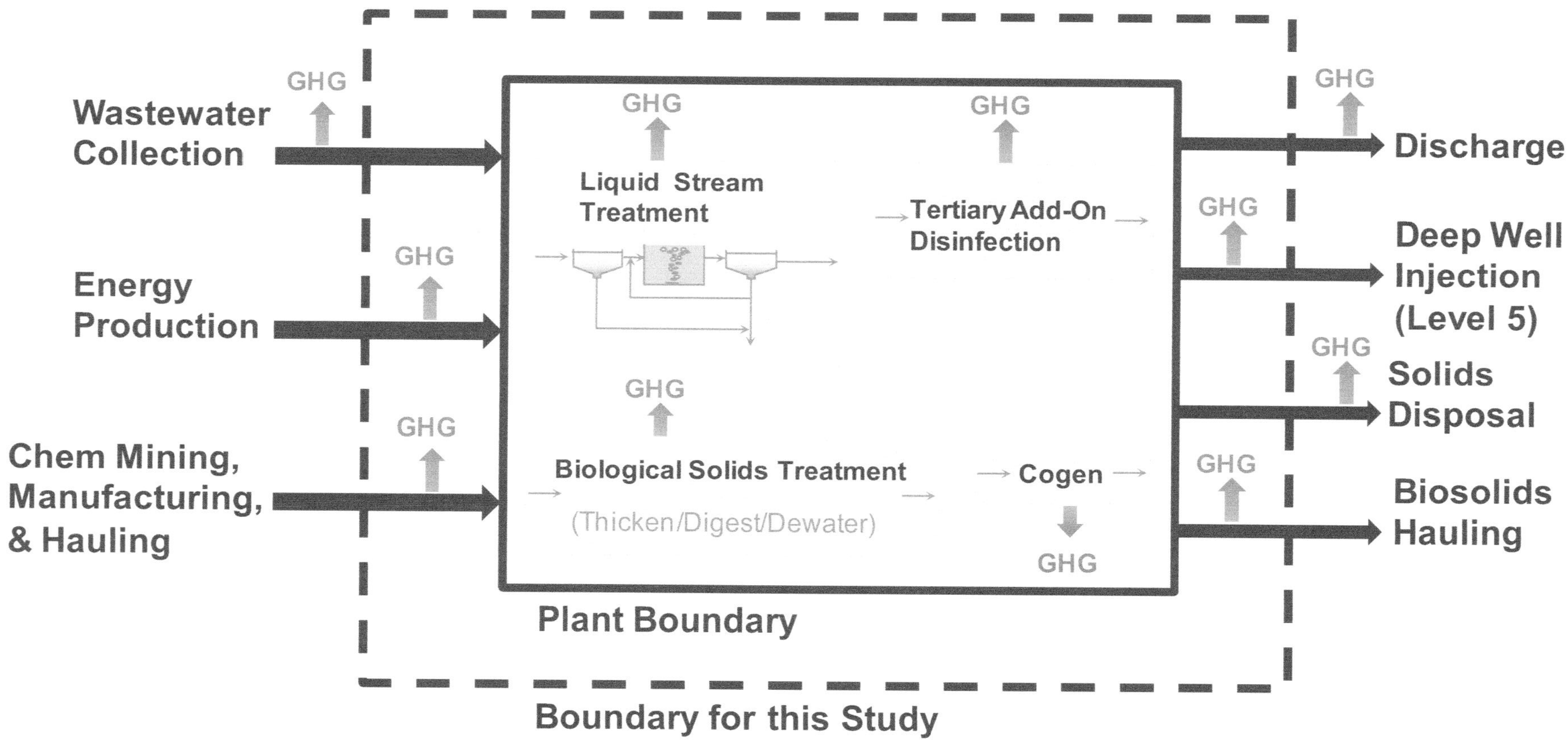

Figure 3-6. Mass Balance System Inputs.

CHAPTER 4.0

SUSTAINABILITY CONSEQUENCES FOR FIVE TIERED LEVELS OF TREATMENT

4.1 Results

The investigation determined that as the degree of nutrient removal increased, the impact on sustainability increased significantly.

4.1.1 Steady State Mass Balance Results

A detailed breakdown of the GHG emissions of each level per unit process/operation in the respective treatment train is provided in Table 4-1. The miscellaneous category includes HVAC, lighting, non-treatment equipment, non-potable water, channel aeration/mixing, chemical feed, and odor control as discussed in Section 3.6. Additionally, the solids values include solids treatment plus the cogeneration biogas values.

Table 4-1. Greenhouse Gas Emission Results.*

Treatment	Level 1 No N/P Removal (CO_2 eq mt/yr)	Level 2 8 mg N/L 1 mg P/L (CO_2 eq mt/yr)	Level 3 4-8 mg N/L 0.1-0.3 mg P/L (CO_2 eq mt/yr)	Level 4 3 mg N/L 0.1 mg P/L (CO_2 eq mt/yr)	Level 5 <2 mg N/L <0.02 mg P/L (CO_2 eq mt/yr)
Primary	240	240	240	240	240
Activated Sludge	2,980	3,820	4,430	4,430	4,430
Tertiary	200	200	650	1,530	6,990
Sub-Total Liquid Stream	3,420	4,260	5,320	6,200	11,660
Solids *	-150	-10	-40	50	-30
Miscellaneous **	1,320	1,320	1,320	1,320	1,320
Total	**4,590**	**5,570**	**6,600**	**7,570**	**12,950**

* The solids treatment stream includes cogeneration of biogas, which results in negative values for most treatment levels

** Miscellaneous results include odor control, non-potable water, high pressure air system, miscellaneous pumping/mixing/sludge storage/transfer pumps, HVAC, administrative buildings energy, and miscellaneous equipment.

The transition from a cBOD mode WWTP (Level 1) to nutrient removal translates to a roughly 20% increase in GHG emissions for Level 2 (4,590 versus 5,570 CO_2 equivalent mt/yr). There is an increase in GHG emissions of 18 and 13% increase from Levels 2 to 3 and from Levels 3 to 4, respectively. The increase in solids GHG emissions under Level 4 is attributed to using the primary solids to fuel the denitrification process from fermentation rather than recovering the energy from the primary solids during anaerobic digestion. Primary solids on average produce more biogas per pound than waste activated sludge which in turn can be used to produce energy at the WWTP. Finally, the GHG emissions for Level 5 reveal a roughly 70%

increase from Level 4 to Level 5. The sharp increase from Level 4 to 5 is based largely on energy demand from MF/RO tertiary treatment plus the energy demand from deep well injection pumping.

Figures 4-1 through 4-3 present values from Table 4-1 as liquid stream GHG emissions (Figure 4-1), the solids stream GHG emissions (Figure 4-2), and the combined GHG emissions (Figure 4-3). The miscellaneous items (HVAC, lighting, non-treatment equipment, non-potable water, channel aeration/mixing, chemical feed, and odor control) fall under the liquid stream treatment (Figure 4-1). The 21% net increase from Level 1 to Level 2 is attributed to the additional aeration required for nitrification (with more efficient aeration as alpha increases from 0.30 to 0.45), MLR pumping, and anaerobic/anoxic zone mixing. The 18% net increase from Level 2 to Level 3 is a result of increased chemical demand (methanol, alum, and polymer). The use of fermentation provides savings in Level 4, but the savings are off-set by the use of methanol in tertiary denitrifying filters and additional chemicals for high HRC. HRC requires more alum and polymer than conventional filtration to remove P. As a result, the net GHG emissions increase is 15% from Level 3 to 4. Finally, the side-stream MF and RO advanced tertiary treatment system requires such elevated pumping levels for treatment and brine management that the GHG emissions increase by 70% while transitioning from Level 4 to 5. It is important to note that the 70% increase of GHG emissions from Level 4 to 5 only remove an additional 1 mg N/L (83 lb N/d for a 10 mgd flow).

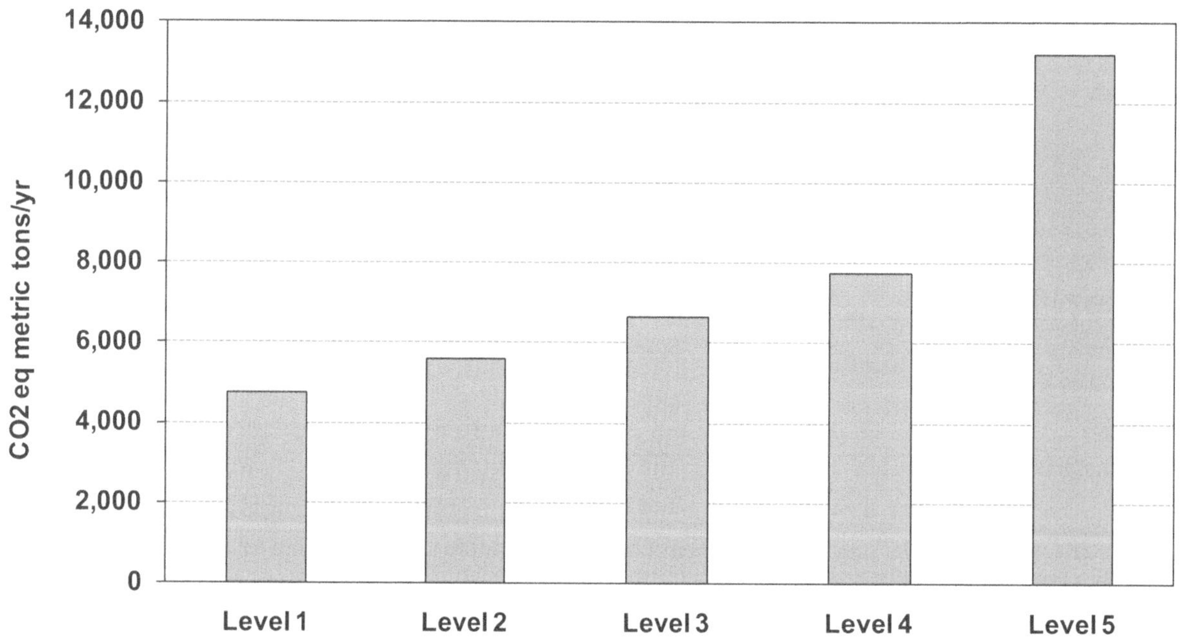

Figure 4-1. Annual GHG Emissions Production Associated with Liquid Stream Treatment.

For all treatment levels, the solids stream values (Figure 4-2) vary based on whether the treatment train removes nutrients and if chemical precipitation is used for P removal. The values include total GHG emissions from solids treatment: pumping, DAFT, anaerobic digestion mixing/heating, dewatering, and chemicals. The lower aerobic SRT associated with Level 1 has a higher yield than Levels 2 to 5. As a result, the increased solids translate to more biogas production. The difference in Levels 2 to 5 relates to chemical usage variability and the use of fermentation in Level 4. The fermentation step sends the sBOD substrate to the activated sludge process rather than using it to create biogas in the anaerobic digestion. Thus, the amount of biogas produced is reduced for Level 4 in the anaerobic digester.

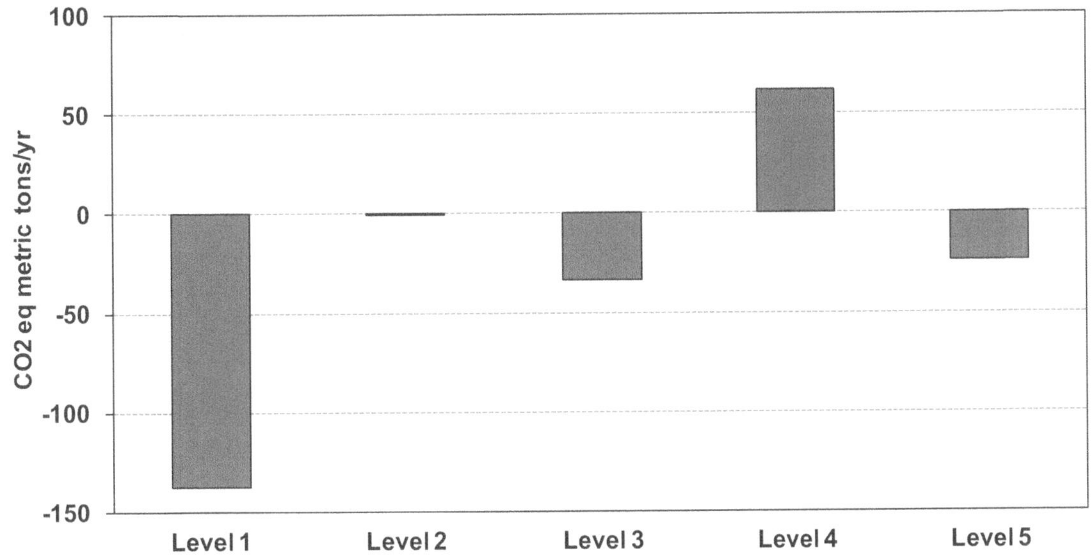

Figure 4-2. Annual GHG Emissions Production Associated with Solids Stream Treatment.

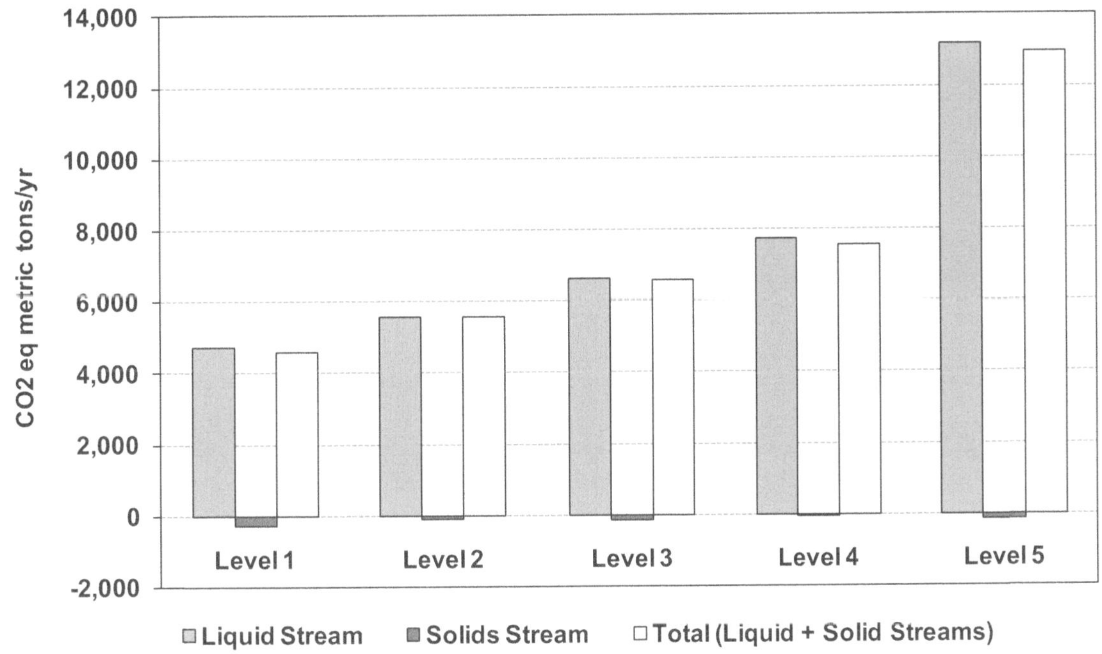

Figure 4-3. Annual GHG Emissions Production Associated with Liquid, Solids, and Combined Streams (Liquid + Solids).

4.1.2 GHG Emissions Distribution per Treatment Level

For each treatment level, the distribution of GHG emissions per treatment level is shown in Figure 4-4. The three largest contributors to GHG emissions are energy (aeration, pumping/mixing, and deep well injection (DWI)), miscellaneous, and chemicals. The aeration assumes an alpha of 0.30 for Level 1 and 0.45 for Levels 2-5. The DWI for Level 5 will vary widely for all the RO reject stream management technologies. By grouping DWI by itself in Figure 4-4, DWI can be replaced with a different technology to determine the overall GHG emissions. Miscellaneous is assumed to be constant, regardless of the treatment level. The impact of chemical production is not significant until Levels 3 to 5. The amount of required chemicals increases with the level of treatment as a means to complement biological limitations. In particular, Levels 4 and 5 both have high rate clarification and tertiary denitrification, which both require methanol and additional alum/polymer.

The only parameter with a range of values is the N_2O emissions (as shown by data range bars). The difference in N_2O emissions between secondary treatment and BNR is negligible. However, the range in values highlights the wider range in values for a plant performing NDN.

The cogeneration values vary for each treatment level. The cogeneration credit is largest for Level 1 since the additional WAS associated with a 2-d aerobic SRT over the 10-d aerobic SRT associated with Levels 2-5 generates more methane.

The biosolids hauling (<100 miles each way) and fugitive methane emissions (IPCC, 2006) were lumped together as there combined distribution is less than 1% of the net GHG emissions.

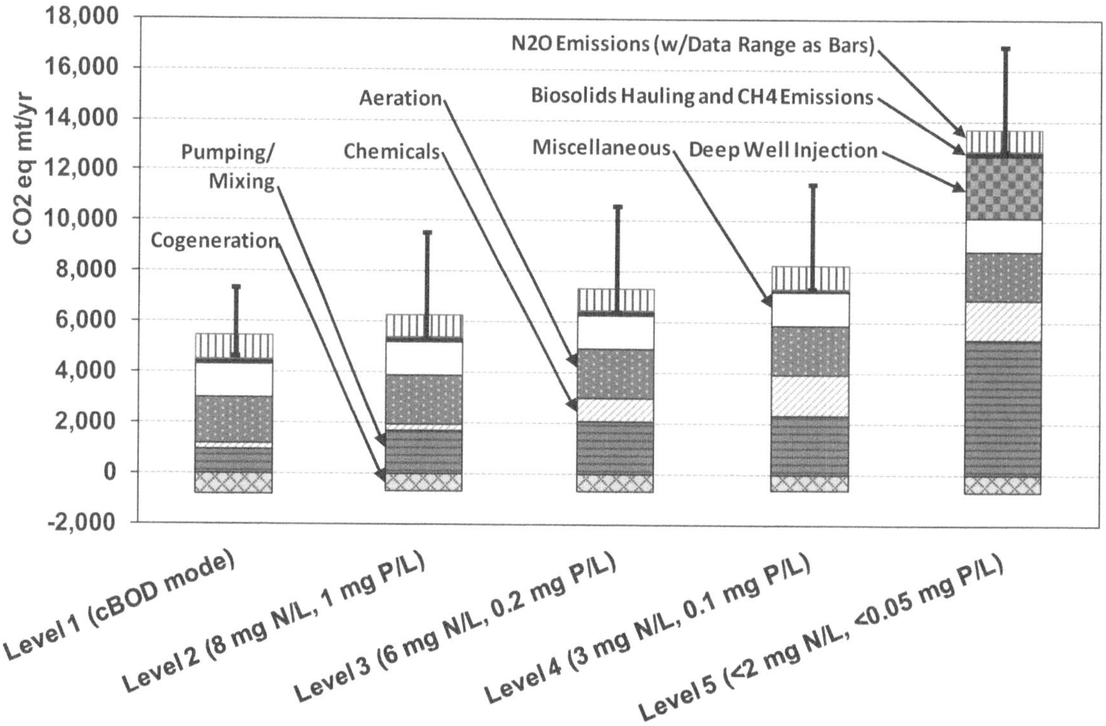

Figure 4-4. GHG Emissions Distribution per Treatment Level.

4.1.3 Incremental GHG Emissions with Increased Treatment

To better understand the sustainability impacts with increased treatment, the incremental GHG emissions increase per additional nutrient load removal was calculated with the transition from one level to the next. The calculation is in accordance with Equation 3 from the Methods Section. This calculation attempts to normalize the data in order to highlight that removing the last few pounds of nutrient is costly in terms of GHG emissions. Figure 4-5 shows the results for the incremental GHG emissions with increased treatment per additional pound of nutrient removed.

The transition from Level 1 to 2 requires three times more GHG emissions per additional pound of N removed. An external carbon source does not impact the transition from Level 1 to 2 because no methanol is required to fuel the process. The incremental increase in GHG emissions becomes more pronounced from Level 2 to 3 as evidenced by the 14 times increase. The increase relates to use of additional chemicals in the form of methanol in activated sludge and alum/polymer in filtration. The trend continues from Level 3 to 4 as it requires 23 times more GHG emissions per additional pound N removed. This increase is attributed to the HRC, HRC chemicals, tertiary denitrification filters, external carbon source, plus the use of fermentation. A high rate clarifier requires chemicals on the order of 50 mg/L alum and 2 mg/L polymer. Additionally, tertiary denitrification filter requires more surface area and thus more energy devoted to backwash. Furthermore, the primary solids that fuel the anaerobic digestion is sacrificed as substrate for denitrification in the main plant flow. Finally, the transition from Level 4 to 5 has an increase of GHG emissions per additional pound of N removed of over 300 times due to the addition of side-stream MF/RO coupled with the energy demand of deep well injection. Based on the incremental GHG emissions increase per additional lb N removed, it can be argued that the point of diminishing returns occurs from Level 4 to 5 as the increase is exponential at 300 times greater GHG emissions per additional lb N removed.

As for P, the incremental GHG emissions increase per additional P load removal plot is comparable to the N findings, but at an overall more pronounced rate. The initial increase in GHG emissions from Level 1 to 2 is the smallest at 0.5 times, which is attributed to only requiring a mixer within the anaerobic zone for biological phosphorus removal. From Level 2 to 3, filtration is required to meet the P discharge requirements. As a result, the incremental increase in GHG emissions per additional pound P removed increases16 times. From Level 3 onwards, the incremental GHG emissions increase per additional pound P removed increases a magnitude with each treatment level. From Level 3 to 4, the increase is 190 times, followed by over 3,000 times from Level 4 to 5. The more pronounced increase for P removal is a result of the removal mechanism being chemically based, along with the P discharge limits being lower. In contrast to the incremental GHG emissions increase per additional N load removal, the P load appears to reach a point of diminishing returns is reached with the transition from Level 3 to 4.

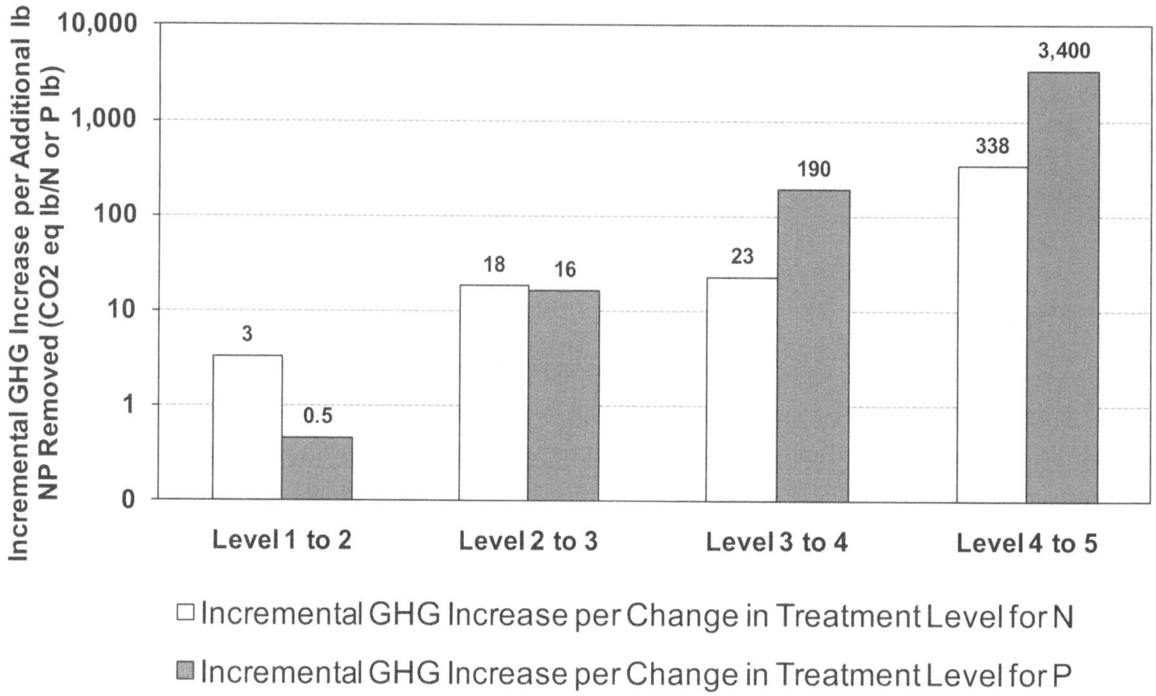

Figure 4-5. Incremental GHG Emissions Increase per Additional lb N or P Removed.

4.1.4 Algal Production Potential in Receiving Waterbodies

Receiving waterbodies are positively impacted by a higher water quality discharge with higher treatment level. The improved water quality associated with a higher treatment level can be evaluated by the surrogate, algae production potential per treatment level. The GHG emissions can then be compared to the benefit in terms of reduction in algal growth. The algae production associated with each treatment level is calculated based on nutrient discharge loads. It was assumed that for every 100 pounds of algae produced requires 7.2 lb of N and 1.0 lb. P (Vallentyne, 1974).

The plots of GHG emissions and algae production per treatment level are shown in Figure 4-6. Nearly 95% of the potential algae production is eliminated when changing from Levels 1 to Level 3 with a 44% increase in GHG emissions. An additional 4% of potential algae production (with respect to Level 1) is eliminated in Levels 4 and 5, while nearly doubling the GHG emissions (6,590 to 12,950 mt CO_2 equivalents/year).

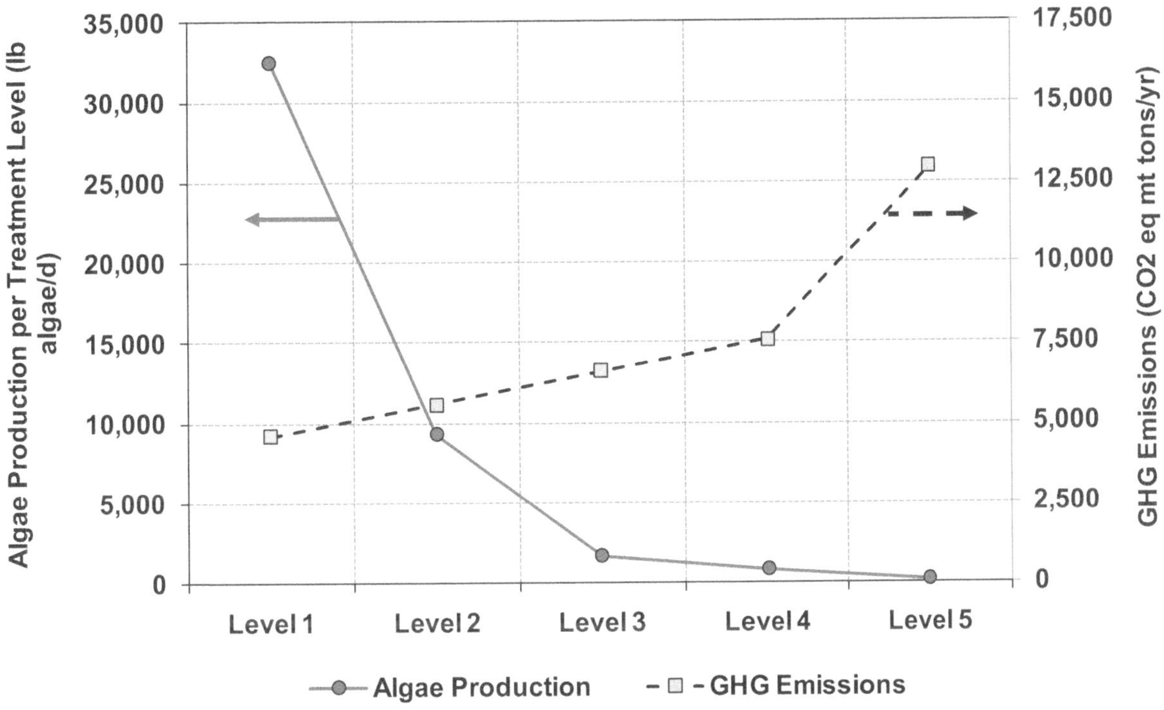

Figure 4-6. GHG Emissions and Algae Production per Treatment Level.

4.1.5 Transportation Impacts on Sustainability for Chemicals

The total plant wide GHG emissions values were calculated for variable chemical hauling distances as shown in Table 4-2. The decision to use chemicals from a regional supplier (750 miles) over a local supplier (100 miles) can impact facilities GHG emissions by 1% (Levels 1, 2, 3, and 5) and 2% (Level 4). This percentage increases to upwards of 5% (Level 4) as the switch is made from local (100 miles) to national (2,000 miles) hauling of chemicals.

Table 4-2. Total Plant Wide Greenhouse Gas Emission for Variable Biosolids and Chemical Hauling Requirements.

Treatment	Units	Level 1	Level 2	Level 3	Level 4	Level 5
		No N/P Removal	8 mg N/L 1 mg P/L	4-8 mg N/L 0.1-0.3 mg P/L	3 mg N/L <0.1 mg P/L	1 mg N/L <0.01 mg P/L
Base Hauling Local (100 miles)	% increase in CO_2 equivalents	--	--	--	--	--
Chemical Hauling Regional (500 miles)	% increase in CO_2 equivalents	1%	1%	1%	2%	1%
Chemical Hauling National (2,000 miles)	% increase in CO_2 equivalents	3%	2%	4%	5%	3%

4.1.6 Costs per Treatment Level

The total project capital, operation, and total present worth project costs are provided in Table 4-3. The total project capital cost is for new Greenfield plants. Total project capital cost includes the construction plus other construction related costs (engineering, construction management, etc). Land costs are not included in the capital cost calculations.

The total project capital cost includes construction, engineering and administration. Base construction calculations were updated to October 2010. The estimates were built process by process with a 25% adder for "undefined structures," 8% for administrative buildings, an additional 28% for site work, yard piping soil conditions, electrical power supply and site electrical.

Contractor costs of 25% were added for field general conditions, mobilization, demobilization, sales tax, overhead, profit, and bonds/insurance. A grand total for construction was calculated and a 20% construction contingency was added. Additional "soft" costs of 28% were added to the construction grand total for planning and design engineering, construction engineering and management, and legal and owner administration.

Table 4-3. Total Project Capital and Operations Costs for a 10 mgd WWTP [i]

Level	Total Project Capital Cost ($/gpd) [ii]	Total Project Capital Cost (Million $) [ii]	Operations Cost ($/MG Treated) [iii]	Operations Cost ($1,000/yr) [iii]	Total Present Worth Project Costs (Million $) [iv]
1 (No N/P Removal)	9.3	93	250	910	110
2 (8 mg N/L; 1 mg P/L)	12.7	127	350	1,260	150
3 (4-8 mg N/L; 0.1-0.3 mg P/L)	14.4	144	640	2,350	180
4 (3 mg N/L; <0.1 mg P/L)	15.3	153	880	3,200	210
5 (1 mg N/L; <0.02 mg P/L)	21.8	218	1,370	4,990 **	300

Notes:

i The total project capital cost are for a Greenfield plant.

ii The total project capital cost are the equipment cost, construction, and "soft costs"

iii Operations cost = energy and chemical cost. Labor and maintenance costs are excluded

iv The assumed discount rate was 5 percent at an escalation rate of 3.5 percent (capital, energy, non-energy)

The total project capital cost highlights that the cost increases with treatment level. The transition from a carbonaceous BOD mode WWTP in Level 1 to a nutrient removal WWTP requires one-third total project capital cost. This can be attributed to an increase in the secondary process (i.e., activated sludge plus secondary clarifiers). The total project capital cost increase from Level 2 to 3 and 3 to 4 are marginal at 13 and 6%, respectively. Level 3 requires filtration and a 5-stage Bardenpho over the A2O used in Level 2. Level 4 requires the addition of primary sludge fermentation, high rate clarifiers, tertiary denitrifying filters, and centrate equalization. The transition from Level 4 to 5 requires an additional 43% increase in total project capital as a result of the side-stream MF/RO treatment.

The discrepancy in values for operations is more pronounced than for total project capital cost. The operations cost increase over 500% from Level 1 to 5. Labor and maintenance activities are not included. There is a steady increase of operations cost between treatment levels, which is largely attributed to increased pumping, mixing, aeration, and chemical demand. The largest increase in operations cost from a magnitude perspective occurs from Level 4 to 5 (56% increase). From a percentage basis, the largest increase occurs from Level 2 to 3 (85% increase) as a direct result of chemical (methanol, alum, and polymer) and filtration requirements.

An attempt to incorporate total project capital and operations costs was done by a total present worth calculation (5% discount rate and 3.5% escalation rate). For each step between Level 1 to 4 requires between 30 and 40 million dollars in total present worth cost. In contrast, the transition from Level 4 to 5 requires 90 million dollars.

In a similar fashion to calculating the incremental GHG emissions increase per additional nutrient load removed (Figure 4-5), the incremental cost per additional pound nutrient removed was calculated as shown in Figure 4-7. This calculation attempts to normalize the data in order to highlight that removing the last few pounds of nutrient is costly for operations and maintenance. Operational cost for N removal increases from 2-9 times from Levels 1 to 2 and Levels 2 to 3, respectively. The incremental cost is relatively constant from Level 3 to 4. As with the incremental GHG emissions increase associated with N removal, the incremental cost increase shows an exponential increase with respect to the other treatments for Levels 4 to 5 (>50 times increase).

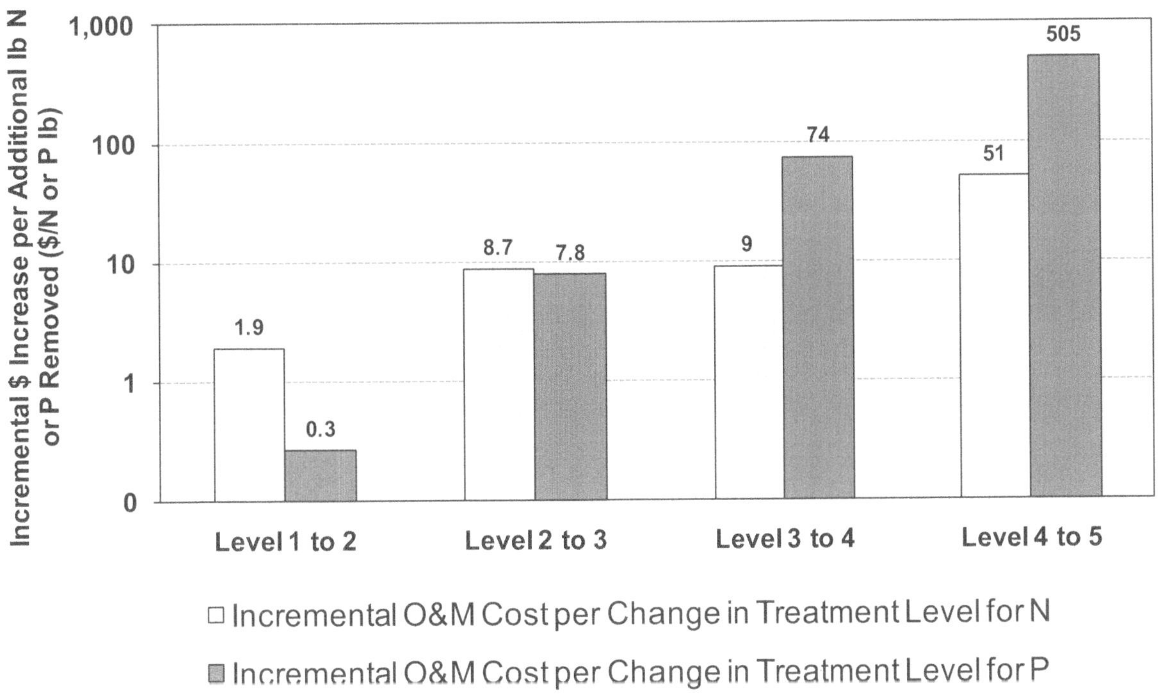

Figure 4-7. Incremental Operations Cost Increase per Additional Pound N or P Load Removed.

The incremental O&M cost increase per additional pound P removed results are in line with the incremental GHG emissions increase (Figure 4-5). The increase with the transition from one treatment level to the next is more pronounced overall than with N. The initial increase is less than 1 at 0.3 times as the only increase in costs is for mixing and pumping for the anaerobic zone to meet the phosphorus objective. From Level 2 onwards, the increase in incremental O&M cost increase per additional pound P removed increases at least 10 times with the transition from one level to the next.

4.2 Discussion

Three areas typically considered in sustainability evaluations are the environmental, economic and social pillars. These three pillars are commonly referred to as the triple bottom line (TBL). The TBL is based on the notion that all three pillars are incorporated into the decision making process to promote a sustainable solution.

4.2.1 Environmental Pillar

Within environmental component, the researchers considered equivalent GHG emissions, reduction in nitrogen and phosphorus load, and potential algal production.

4.2.1.1 Which Parameter is Governing GHG Emissions?

Based on the overall GHG emissions for each treatment level (Figure 4-4), a ranking of the average GHG emissions distribution for all treatment levels is as follows: energy (pumping, mixing, aeration, DWI) > miscellaneous > chemicals > N_2O emissions > cogeneration > biosolids hauling and CH_4 emissions. Regardless of treatment level, energy is the dominant parameter governing GHG emissions. Energy accounts for between 55-65% of GHG emissions for all treatment levels.

Energy as the primary GHG emissions culprit is in agreement with Keller and Hartley (2003) where they compared a municipality with liquid stream anaerobic treatment versus the layout of Level 1 (Figure 3-1). They found a municipality based on liquid stream anaerobic treatment contributes 1.0 kg CO_2/kg COD removed versus 2.4 kg CO_2/kg COD removed for a cBOD plant (Level 1). They attributed the 240% increase for the cBOD WWTP on the additional energy required for pumping, mixing, and aeration. Research by Cakir and Stenstrom (2005) advanced the comparison of anaerobic versus aerobic treatment for treating the wastewater. Their study considered different influent ultimate BOD (uBOD) concentrations and found that if the uBOD levels were greater than 700 mg/L, anaerobic treatment was the preferred WWTP configuration from a GHG emissions and stable process perspective.

Tripathi (2007) also considered different types of treatment plants: a drinking water plant, a WWTP, and a WWTP with reclamation. For all three facilities, energy demand was the primary contributor to GHG emissions that ranged from 44-91% of the total GHG emissions. A study by Racoviceanu et al. (2007) considered energy demand for a drinking water treatment plant, where on-site pumping alone represented 90% of GHG emissions. Strategies to reduce demand consists of i) adjustment in treatment plant configuration as suggested by Keller and Hartley (2003), ii) more efficient on-site pumping/aeration/mixing equipment, or iii) a reduction in wastewater source load.

The third largest GHG emissions contributor in Figure 4-4, chemicals (ranged from 4 to 21% of total GHG emissions), becomes increasingly important beginning at Level 3. The chemicals are necessary to serve as an external carbon source for denitrification (methanol) and a metal salt (alum) to precipitate P out of solution. For treatment levels requiring methanol (Levels 3, 4, and 5), the methanol represents between 4-9% of total GHG emissions (about 600 CO_2 eq mt/yr). The energy requirements for methanol are 250 kWh/MG treated, which is less than half of the 600 kW/MG treated used by Kang et al. (2009). The disparity in values with Kang relates to dose as they assumed more than double the values used in this study.

The N_2O emissions are highly variable as indicated by the data range bars for non-BNR and BNR configurations. The rationale behind using Ahn et al. (2010b) relates to the fact that the values are from a survey of several WWTPs rather than a single WWTP. This database is based on real-time samples using state of the art analytical equipment. At the peak data range levels, nitrous oxide emissions would represent the second most important parameter behind energy. Given the potential role in overall GHG emissions and its potency as a GHG (296 times more potent than CO_2), research on the mechanisms of the formation of N_2O is being pursued by several groups to better understand their role at WWTPs.

The hauling associated with manufacturing and delivering metal salts, such as alum or ferric is most likely understated in the calculations. For example, alum was used and a 100 mile hauling distance assumed. Although the calculation used the GHG emissions associated with alum manufacturing, the GHG emissions multiplier does not include the transport from source to manufacturing. For example, Jamaica is the nearest source of the material (i.e., bauxite) used to manufacture alumina. Following alumina manufacturing in Jamaica, it is shipped to the U.S. and further processed to produce alum.

Although the GHG emissions calculation incorporates the energy and materials associated with mining to manufacturing a final product, it does not include transport from mining to manufacturing. Additionally, other environmental impacts exist that are excluded. In the case of the bauxite process, the red mud tailings associated with the process create an array of environmental impacts unaccounted for when using the GHG emissions index (Coke et al., 1987). As a result, the GHG emissions multiplier used for alum is most likely understated as it does not completely capture the overall transport (mine to manufacturing) or the environmental impacts associated with mining alum.

The smallest contributor, fugitive methane gas, might be understated due to a lack of consensus on accounting. Emissions of fugitive methane (23 times more potent than CO_2) from raw sewage might be significant as indicated earlier in this document which, in turn, would reduce the GHG emissions load proportion from energy. A more detailed discussion on methane and nitrous oxide can be found in Sections 4.2.1.3 and 4.2.1.5, respectively.

4.2.1.2 Plant Configurations

The GHG emissions and cost associated with the treatment level are dependent on the selected plant configuration. Other plant configurations could be implemented to meet treatment levels. A change in the plant configuration would impact both GHG emissions as well as cost. Research by Hunt et al. (2009) considered the GHG emissions generated when using different nutrient removal configurations (Johannesburg, MLE, Step-BNR, Step-Feed, and Multi-Sludge) to meet a defined treatment level (TN $\leq$ 10 mg N/L and TP $\leq$ 1 mg P/L). Their efforts considered both operations and construction for GHG emissions. The results found Step-BNR < Step-Feed < Johannesburg = MLE < Multi-Sludge. The emissions ranged from 26,000 to 42,000 CO_2 eq mt which is attributed to variable pumping, mixing, chemical requirements and basin volume requirements.

Newer or emerging technologies could impact the GHG emissions. The use of RO technology to meet Level 5 increases GHG emissions significantly, not only because the technology is energy intensive, but the disposal of brine adds a significant amount of GHG emissions. If NF emerges as a viable technology to achieve the discharge objective (<2 mg N/L; <0.02 mg P/L) its use would greatly reduce the GHG emissions because a NF membrane would allow mono-valent ions to pass through the membrane and in turn have a less concentrated reject stream. As noted in Section 1.9.4.6, research by Xu et al. (2010) found drinking water plants with NF membranes have the ability to remove DON while allowing the ionic nitrogen species (e.g., ammonium) to pass through the membrane. The NF reject stream is lower than RO and also less concentrated which in turn would reduce reject stream management costs.

4.2.1.3 Ancillary Benefits of a Longer SRT

In addition to the quantified GHG emissions and algal production plots related to increasing treatment level, there are several ancillary benefits associated with transitioning from secondary treatment to nutrient removal as follows:

- Lower BOD/TSS discharge load
- Higher removal of TOrCs (Horz et al., 2004)
- Higher removal of heavy metals (Stasinakis et al., 2003)
- Improved water quality for downstream users increases water use flexibility, in particular Level 5
- Less algal blooms as there is less nutrient load discharged
- Reduced receiving water dissolved oxygen demand due to ammonia removal
- Reduced ammonia discharge loads, which is toxic to several aquatic species
- Improved water quality for habitat, especially as it relates to biodiversity and eutrophication (Beck et al., In press)
- Secondary clarifier effluent more conditioned for filtration and disinfection (Leu et al., In press; Chan et al.; In press)
- Greater process stability from the anaerobic/anoxic zones serving as a selectors (Jenkins et al., 2004)

When performing a TBL on a specific plant, it is critical that the items listed above be factored into the decision making process. Although most of the listed items fall under the environmental pillar, several of them stretch across multiple pillars. For example, improved water quality for downstream users impacts all three pillars.

4.2.1.4 Methane Emissions

Although this study excludes sewer system conveyance, it is important to acknowledge the potential impact of methane emissions formation during this process. Although largely ignored at WWTPs, methane formation is gaining recognition as a significant GHG emission contributor during the anaerobic conditions commonly encountered during sewer system conveyance. Atmospheric scientists point towards domestic sewage conveyance as a major anthropogenic methane source (Minami and Takata, 1997).

Methane sewer field measurements, coupled with controlled experiments, revealed that hydraulic residence time (HRT) in the sewer collection system and pumping frequency along the sewer system govern methane production (Guisasola et al., 2008; Guisasola et al., 2009). Given the variability in these two operational parameters, HRT and pumping frequency, it is difficult to develop default values for methane production. Rather, a range of values is a more reasonable and valid approach. These findings suggest that sewer methane emissions range from 20 to 100 mg/L as COD. Given the potency of methane as a CO_2 gas equivalent, the emissions indicate that collection system methane could be equivalent to 15-65% of the CO_2 equivalents of the electricity used to treat the same amount of COD at the wastewater treatment plant.

These recent revelations by Guisasola et al. (2008; 2009) on the magnitude of methane production in sewer collection systems should prompt future research to better understand formation mechanisms in collection systems. This is especially critical from a GHG emissions perspective as the methane multiplication factor for CO_2 equivalents is 23 (IPCC, 2006).

4.2.1.5 Nitrous Oxide Emissions

As previously stated, the IPCC (2006) and U.S. EPA (2009) approaches estimate GHG emissions using flow based emission factors for N_2O emissions that only consider N_2O formation to occur during denitrification. This approach has two inherent short-comings in that a flow based approach does not account for variable wastewater strength coupled with the fact that N_2O emissions occurs during both nitrification and denitrification (Yu et al., 2010). Based on these short-comings, this study used the U.S. national survey of N_2O emissions compiled by Ahn et al. (2010b). Their study investigated 12 different plants nationwide (3 non-BNR, 8 BNR, and 1 sidestream treatment). The average values for non-BNR (Level 1) and BNR (Levels 2 to 5) were used in the study. Given the fact that the treatment technology schematics used to meet the treatment objectives in this study could be different, the average values from Ahn et al. (2010b) were used in this report. The wide variability in N_2O emissions was highlighted by providing error bars of the standard deviation from the 12 different plants monitored for N_2O emissions by off-gas testing.

The known mechanisms associated with N_2O emissions were used to develop a description of N_2O emissions and reduction strategies for all five treatment levels (Appendix B). WWTPs operated under cBOD mode (Level 1) have the ability to potentially nitrify part or all of their influent TKN load at high temperatures (e.g., during summer). Under such circumstances, N_2O emissions are expected to be especially high due to incomplete nitrification (incomplete ammonia, NH_3 conversion to either nitrite, NO_2^- or nitrate, NO_3^-) or during periods of high NO_2^- accumulation.

Intermittent nitrification (where nitrification occurs sporadically from the imposition of excess influent TKN loads) can lead to N_2O emissions. Because these loads are not adequately buffered at low operating SRT (<2 d), the diurnal peaks in influent TKN loading can lead to especially high NH_3 accumulation in such plants (Ahn et al., 2010a). Given that N_2O emissions from nitrification are correlated with the extent of NH_3 accumulation, this is yet another factor exacerbating N_2O emissions from plants not designed and operated for BNR, but where intermittent nitrification can occur (Ahn et al., 2010a; Yu et al., 2010).

The N_2O emissions from a WWTP designed and operated to remove nitrogen are expected to be lower than for a cBOD removing WWTP that unintentionally nitrifies. This reduction is primarily because of the higher operating SRT in a nutrient removal WWTP. As temperatures rise in a WWTP that operates in cBOD mode, the nitrification growth kinetics increase and the process can enter in and out of nitrification. This dynamic situation can lead to increased N_2O emissions. Additionally, the NH_3 and NO_2^- concentrations in a nitrifying WWTP with high degrees of MLR are also lower than a cBOD mode WWTP, which leads to a reduction in nitrification and denitrification associated N_2O emissions.

There are two specific strategies for optimizing MLR that could minimize the overall N_2O emissions potential for a given effluent TN concentration of less than 8 mg-N/L. The first strategy is to achieve near complete nitrification, so as to minimize the NH_3 concentrations in the system, which implies that NO_3^- will constitute the majority of the effluent TN load.

The second strategy is to minimize the extent of plug-flow in the system, especially in the aerobic zone and thus minimize the NH_3 gradient therein (Ahn et al, 2010a; Yu et al., 2010). Given the positive correlation between NH_3 concentrations and N_2O emissions, minimizing plug-flow conditions can also minimize the magnitude of N_2O emissions especially in the upstream regions of the aerobic zone.

It is also conceivable that COD limitation in the anoxic zones can result in N_2O emissions as documented in some studies previously (Hanaki et al., 1992). However, through the intensive monitoring campaigns conducted as part of a WERF study, Ahn et al. (2010a) found that the N_2O emissions in anoxic zones have been minimal across the board, both in the presence and absence of external COD addition. Recent lab-scale results also point to the fact that transient COD limitation contributes minimally to N_2O emissions (Lu and Chandran, 2010).

As more studies are carried out and more data is generated, a greater understanding on the role of feed delivery, configuration, and operation of BNR plants on overall N_2O emissions is emerging. This will allow for a more accurate GHG emissions accounting of N_2O emissions.

4.2.1.6 Incremental GHG Emissions Increase per Treatment Level

The use of Equation 3 to calculate the incremental GHG emissions increase per additional nutrient load removal with a transition from one level to the next attempted to highlight that removing the last few pounds of nutrient is costly in terms of GHG emissions. Figure 4-5 illustrated this point based on the exponential increases that suggest that a point of diminishing returns is reached for both nitrogen and phosphorus. For nitrogen, moving beyond Level 4 treatment generates significant increases in GHG emissions while achieving a small reduction in nitrogen load. From Level 4 to 5, the average discharge nitrogen concentration reduces from 3 to 2 mg N/L (additional 83 lb/d N removed for the 10 mgd plant).

The incremental GHG emissions values for phosphorus (Figure 4-5) are even more pronounced than for nitrogen. The increase in GHG emissions reaches a tipping point from Level 3 onwards. Moving beyond Level 3 results in substantial GHG emission increases for a relatively small removal of phosphorus (additional 8.3 lb P/d from Levels 3 to 4). Rather than attempt to comply with a Level 4 or 5 treatment objective, a combination of Level 3 treatment objectives and best management practices for non-point source loads might be a more viable approach at meeting the overall load targets on the receiving waterbody.

The algae production plot (Figure 4-6) further supports the finding that relatively small nutrient load savings are achieved as the level of treatment increase from Level 3 to 4 to 5. A point of diminishing returns to reduce algal production is apparent following Level 3.

4.2.1.7 Dissolved Organic Nitrogen

As discussed in Section 1.9.2, rDON is the limiting factor in achieving low N objectives ($N < 3$ mg N/L). rDON effluent concentrations have been compiled as shown in Figure 4-8 and 4-9. Figure 4-8 is based on a compilation of rDON effluent values nationwide, with particular emphasis on the Chesapeake Bay and Florida. The 50th percentile value is 1.2 mg N/L, which suggests that more than half the facilities cannot reliably meet the 1 mg N/L objective of Level 5. The second figure (Figure 4-9) focuses on Maryland and Virginia municipalities that discharge to the Chesapeake Bay. Over 65% of the municipalities have rDON discharge values of less than 1 mg N/L. These two plots illustrate the variability of rDON discharge values which can directly impact a municipality's ability to reliably meet stringent discharge values. Effluent limits that require nitrogen values of 2 mg N/L or less could dictate the use of expensive and energy intensive advanced membrane technologies to achieve compliance.

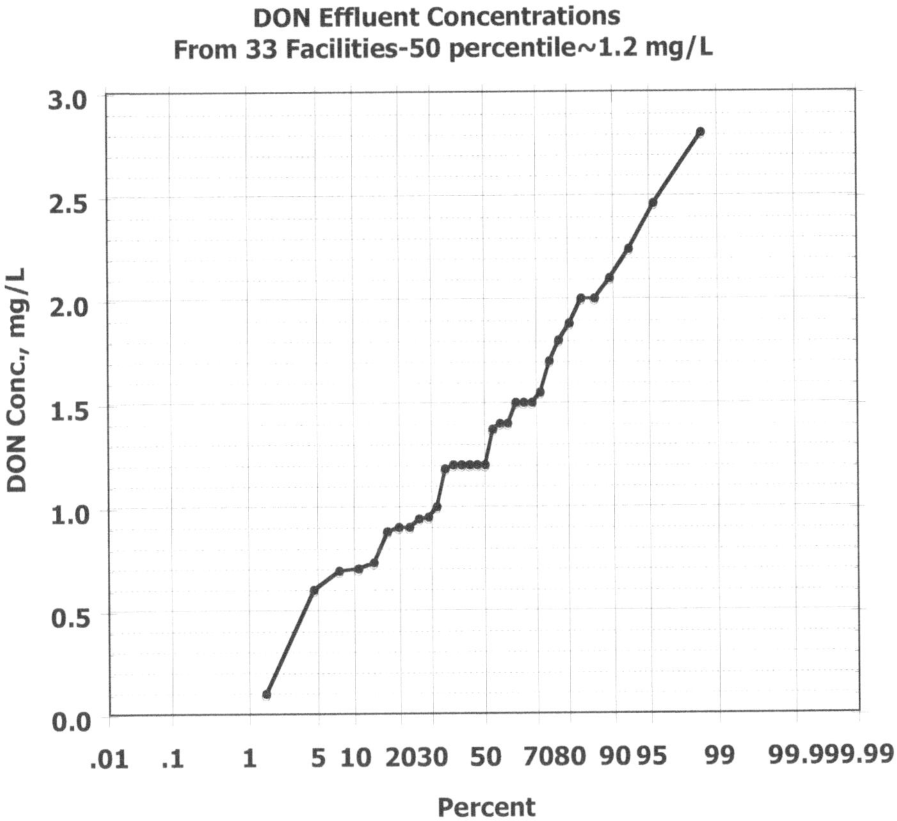

Figure 4-8. rDON Effluent Concentrations from 33 Facilities. (Stensel et al., 2008)

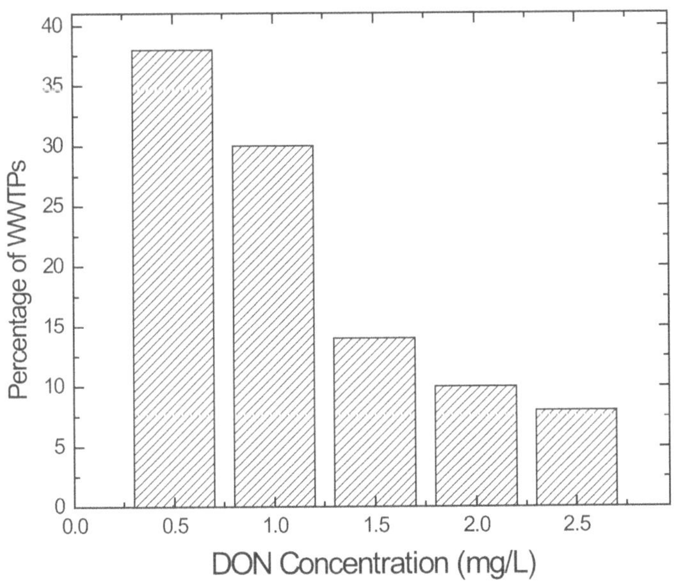

Figure 4-9. Effluent DON Distribution in Maryland and Virginia Plants. (Pagilla, 2007)

The primary permitting issues with rDON relate to bioavailability and whether or not to exclude it from the discharge N permit limits. Given the inability to remove rDON with current technologies during wastewater treatment, the impact of the rDON species leaving a plant could be costly. A high rDON species would require the effluent inorganic nitrogen be reduced to very low values to meet stringent limits.

The recalcitrance of DON over a two-week study at the Truckee Meadows Water Reclamation Facility was assessed by Pehlivanoglu and Sedlak (2004). They found that algae by itself took up 10% of the effluent DON in the absence of bacteria, whereas algae combined with bacteria facilitated a 60% removal of effluent DON. A symbiotic relationship appears to exist, whereby bacteria initiate DON breakdown provided by algae. Their findings suggest that effluent DON can contribute to eutrophication. This is especially the case for Truckee Meadows, where DON accounts for ~85% of discharged TN.

Biological removal mechanisms of rDON in the secondary process are still largely unknown. The DON entering is typically larger (high molecular weight) molecules than those present in the effluent. Roughly 50% of the soluble DON can be removed through adsorption, such as granular activated carbon (Keller et al., 1978; Gur-Reznik et al., 2008). It is anticipated that effluent DON contains elevated levels of colloidal amino sugars (from cell walls; DCAA) (Esparza-Soto et al., 2003; Westerhoff and Esparza-Soto, 2003). Being colloidal, one would expect that it can be removed by chemical coagulation, followed by filtration and/or sedimentation. Randtke and McCarty (1979) considered lime, ferric, alum, and others and found removal efficiencies for Ferric>Lime>Alum with a range of 28-40% DON removal. More recent research by Arnaldos and Pagilla (2010) focused on simultaneous rDON and rDOP removal by coagulation / microfiltration using alum found 69 and 72% removal, respectively.

AOPs and selective membranes with larger size exclusion than RO represent another potential removal technology. Dwyer et al. (2008a, b) investigated the viability of the AOP process, UV/H_2O_2, at rDON removal. To remove roughly 25% of the rDON requires H_2O_2 concentrations on the order of 3,000 mg/L. Research on NF membranes have found on the order of roughly 80% rDON rejection (Xu et al., 2010). Although both AOP and NF membranes might serve as a future strategy for rDON removal, the currently industry accepted standard is RO, as used in this report.

4.2.1.8 Dissolved Organic Phosphorus Removal

Research on TP removal is lacking compared to what's available for TN. The lack of knowledge on TP removal relates to the fact that P permits have historically been less difficult to meet from a treatment perspective. As the permits become more stringent, research will need to better characterize the P species so that engineers can better design removal strategies.

As with nitrogen, the rDOP species becomes increasingly important for meeting the treatment objectives for Level 5. Although RO can reliably meet the Level 5 phosphorus objective (<0.02 mg P/L), the incremental GHG emissions increase for the additional 0.02 mg P/L phosphorus removed is on the order of 994 times greater than Level 4. As with rDON, there is wide variability in rDOP exiting a plant. The impact of the rDOP species leaving a plant could be costly.

Recent research is advancing our understanding on both P species and their removal mechanisms. Liu et al. (2010) investigated P removal for various phosphorus species in BNRs and chemical P removal WWTPs. The results suggest that BNRs effectively remove most TP species, with higher removal efficiencies (>93%) that included soluble reactive P (sRP), particulate reactive P portion (pRP), and particulate acid hydrolysable P (pAHP). In contrast, the organic P species had lower removal efficiency (78%). The species least effectively removed was soluble acid hydrolysable P (sAHP) (<40%). Chemical P removal process was more effective for elimination of sRP, sAHP and particulate organic P portion (pOP), but was not as effective for removing pAHP and, it exhibited nearly no removal of DOP.

This ongoing research will lead to a better understanding of the species and mechanisms associated with removal. Thus an opportunity might arise to meet treatment Level 5 without the use of RO and subsequently greatly reduce GHG emissions.

4.2.2 Economic Pillar

If a WWTP was retrofitted from secondary to BNR treatment, the total project capital cost is expected to be lower and site specific than provided in Table 4-3. The total project capital cost for retrofits will depend on the available assets (e.g., tankage), layout, technologies in place, age of the WWTP, available space, climate, and other site specific factors. The amount of available tankage, piping layout, blower capacity, etc. is case specific and governs the cost associated with a retrofit. Additionally, because the required aerobic SRT for full nitrification is typically governed by temperature, a WWTP located in a warm climate will require a smaller increase in SRT than facilities located in cold climates. A culmination of these factors illustrates the variability in total project capital cost for facilities that undergo retrofits.

4.2.3 Social Pillar

Rather than focusing solely on the economic impacts at the treatment plant under the economic pillar, the community may include the social implications of the process in the assessment of sustainability. This component was not quantified because it is heavily case specific and reflects the values of the particular community. Public officials in the community can work along with stakeholders to determine values and priorities in meeting permit requirements. For example, the value of water might be increased if a downstream user that provides jobs requires a higher level of treatment than the permit level.

Besides the social costs, the social pillar includes other factors that are case specific, such as stake-holder participation, safety, jobs for on-site and downstream industry, security, monetizing the cost of reclaimed water, etc. Recent publications by Guest et al. (2009; 2010) provide guidelines to incorporate the TBL's social pillar.

The largest challenge associated with the social pillar is quantifying individual components (e.g., jobs). Most available social metrics only consider the public safety risk assessment (Hunkeler et al., 2006). Social metrics must be developed to include the non-risk aspects. Even if the social metrics existed, the remaining challenge is to incorporate and present them to the stakeholders.

Two such visual approaches exist that serve as a starting point in stakeholder participation: force field diagram (Daigger, 2009) and causal loop diagrams (Wolstenholme, 1999). For both approaches, the intention is to provide the stake-holders with a visual approach that highlights their decision making impacts. There is a need to advance the TBL's social pillar component.

4.2.4 Point versus Non-Point Loadings

Removing nutrients uses significant energy and chemicals that produce GHG emissions. Reaching exceptionally high levels of nutrient removal results in such high GHG emissions per incremental pound of N and P removed that a point of diminishing returns is reached. Achieving exceptionally high levels of nutrient removal is exacerbated by the presence of recalcitrant dissolved organic nitrogen and phosphorus, for which economical removal mechanisms have not been developed. These recalcitrant compounds may not impact water quality significantly in receiving waters. Therefore, removing nitrogen to exceptionally low levels may require removal of virtually all nitrate, nitrite, and ammonia as well as rDON, which can only be done with Level 5 treatment. Level 5 treatment also requires removing particulate phosphorus and rDOP.

Nonpoint sources of nitrogen and phosphorus are significant (Chapter 2.0). A significant reduction in nitrogen and phosphorus might be achievable by using sustainable best management practices that could have considerably less impact on GHG emissions than high levels of treatment at municipalities.

A rational and holistic approach to address both point and nonpoint discharges of N and P could result in a win-win situation: reduced nutrient loads to our waterways with manageable impacts on GHG emissions production.

CHAPTER 5.0

SUMMARY, FUTURE WORK, AND CONCLUSIONS

5.1 Summary

The wastewater industry has seen increasingly tighter restrictions on nutrient discharges, primarily nitrogen and phosphorus. This report provides a bench-top analysis on finding the balance between nutrient removal and sustainability. The objective is to determine if a point of diminishing returns is reached where the sustainability impacts of achieving increased levels of nutrient removal outweigh the benefits of better water quality.

Three areas typically considered in sustainability evaluations are the environmental, economic and social pillars. These three pillars are commonly referred to as the triple bottom line (TBL). The TBL is based on the notion that all three pillars are incorporated into the decision making process to promote a sustainable solution.

For this investigation, analyses were carried out on the environmental and economic components of the TBL. Within environmental component, the researchers considered equivalent GHG emissions, reduction in nitrogen and phosphorus load, and potential algal production. The economic pillar included capital and operations costs. Because the social pillar is case specific, the researchers did not carry out any analyses. Instead, the social pillar was discussed but not quantified. Overall, this is a simplified approach that provides consistent methodology for measurement without investigating the full array of economic, social and environmental sustainability impacts.

The measurement used to quantitatively measure the environmental impacts is equivalent GHG emissions since it provides a means to normalize data for comparative purposes. Given that sustainability is a broad term, the focus was placed on the following variables for the analysis:

- GHG emissions
- Capital costs
- Operating costs
- Energy demand
- Air and water quality
- Consumables, such as chemicals, gas, diesel, etc.

This study compared five different treatment levels of nutrient removal as shown in Table 5-1. With each treatment level, a mass balance was developed to quantitatively determine the GHG emissions equivalents based on the aforementioned parameters.

Table 5-1. Treatment Level Objectives.

Level	Biochemical Oxygen Demand (mg/L)	Total Suspended Solids (mg/L)	Total Nitrogen as Nitrogen (mg N/L)	Total Phosphorus as Phosphorus (mg P/L)
1	30	30	-	-
2	<30	<30	8	1
3	<30	<30	4-8	0.1-0.3
4	<30	<30	3	0.1
5	<30	<30	<2	<0.02

The investigation determined that as level of treatment for nutrient removal increased, the impact on sustainability (as measured by green house gas) increased significantly. Scope 1, 2, and 3 GHG emissions are included with the system inputs. Achieving treatment Level 5 (N<2 and P<0.02) resulted in a dramatic increase of GHG emissions over treatment Level 4 (more than doubling). It appears that a point of diminishing returns may be reached by achieving a small increase in potential water quality that results in a very large impact on the overall environment (as measured by GHG emissions). For the purpose of quantifying and compare the GHG emissions and costs, a treatment plant were evaluated for all treatment levels at a nominal flow of 10 mgd.

5.2 Conclusions

1. Removing N and P to treatment Levels 4 and 5 results in negative sustainability impacts with a corresponding small improvement to water quality.
2. GHG emissions increase significantly when nutrient removal is incorporated in a wastewater treatment plant. For a nominal 10 mgd WWTP, the GHG emissions increased from 4,590 CO_2 equivalent metric tons per year for a cBOD mode WWTP (Level 1) to 7,570 CO_2 equivalent metric tons per year, where nitrogen and phosphorus removal requirements are 3 mg N/L and 0.1mg P/L (Level 4). Further reductions in N and P to <2 mg N/L and <0.02 mg P/L (Level 5) increase the GHG emissions to 12,950 CO_2 equivalent metric tons per year.
3. The level of nutrient removal also increases the total project capital, operating, and total present worth costs. The total project capital cost increased from $93 million for a Level 1 WWTP to $153 million for a Level 4 WWTP and $218 million for a Level 5 WWTP. The difference in operating cost between treatment levels is more pronounced than total project capital cost (Level 1 are $910,000 per year and $3,200,000 per year for a Level 4; more than three times as much). The incorporation of total project capital and operating costs into a total present worth cost illustrates the overall difference in cost between treatment levels, where Level 4 is nearly twice as much as Level 1.
4. The incremental GHG emissions increase in lbs CO_2 eq per lb nutrient removed increases exponentially as the level of treatment increases. The incremental values from Level 1 to Level 2 are 3 for N and 0.5 for P. The intensity for moving from Levels 4 to 5 is 338 for N and 3,400 for P.

5. Recalcitrant dissolved organic nitrogen, commonly referred to as refractory dissolved organic nitrogen (rDON), impairs a WWTPs ability to reliably achieve low TN objectives. Effluent limits that require nitrogen values of 2 mg N/L or less might require the use of expensive and energy intensive strategies, such as reverse osmosis, that result in elevated GHG emissions.
6. As with rDON, recalcitrant dissolved organic phosphorus, commonly referred to as rDOP, hinders a WWTPs ability to meet the total phosphorus value required of Level 5. As a result, an energy intensive technology, such as reverse osmosis (RO), is required to ensure reliably meeting the treatment objective.
7. The GHG emissions for all treatment levels are dominated by energy demand.
8. Based on the incremental GHG emissions increase per additional pound nutrient removed, a point of diminishing returns appears to be reached between Levels 3 and Level 4 for nitrogen removal, and between Level 4 to 5 for phosphorus removal.
9. Using reverse osmosis to achieve extremely low levels of nitrogen and phosphorus increases costs and GHG emissions. Brine reject management remains a challenge for reverse osmosis applications, especially for inland applications.
10. Ancillary benefits associated with a longer sludge age in nutrient removal treatment levels exist and should be accounted for in a full evaluation (e.g., secondary clarifier effluent more conditioned for filtration/disinfection, less biosolids produced, etc.).

5.3 Future Work

1. Developing a rational approach to receiving water nutrient loading, such as requiring "best practices" for non-point sources, may be much more beneficial than requiring wastewater treatment plants (WWTPs) to remove nutrients to levels that adversely impact sustainability. Further dialog with regulators on a national and local level is needed to find a regulatory framework that best protects water quality and manages GHG emissions using both non-point and point source control.
2. Clearly define whether N or P is the limiting nutrient for specific receiving waters.
3. Incorporate bioavailability of N and P in models and assessments to set permit limits. Consider the impact of rDON and rDOP and the slow transformation kinetics of these compounds in nature.
4. Determine the role of peaking factors in GHG emissions per capita for the different treatment levels, particularly in combined sewer plants (majority in United States). Oversized and underutilized facilities translate to higher GHG emissions per capita with increased levels of treatment.
5. Continue work on finding more sustainable technologies for removing nitrogen and phosphorus. Examples include using anammox bacteria to remove ammonia and recycling chemical sludge for phosphorus removal.
6. Initiate discussion with the U.S. EPA regarding the impact of refractory nitrogen and phosphorus on water quality and impact on treatment costs and sustainability.

Determine the role of coagulant chemical selection in removing refractory N and P.

APPENDIX A

A DETAILED BREAKDOWN OF GHG EMISSIONS

Table A- 1. Detailed Breakdown of Greenhouse Gas Emission Results.

Treatment	Level 1 No N/P Removal (CO_2 eq mt/yr)	Level 2 8 mg N/L 1 mg P/L (CO_2 eq mt/yr)	Level 3 4-8 mg N/L 0.1-0.3 mg P/L (CO_2 eq mt/yr)	Level 4 3 mg N/L 0.1 mg P/L (CO_2 eq mt/yr)	Level 5 <2 mg N/L <0.02 mg P/L (CO_2 eq mt/yr)
Liquid Stream					
Influent Pumping	209	209	209	209	209
Primary Treatment	28	28	28	28	28
Sub-Total Pri.	240	240	240	240	240
Anaerobic Mixing		67	81	78	78
Anoxic Mixing		153	184	178	178
MeOH Activated Sludge			562	574	575
Aeration	1,804	1,949	1,949	1,949	1,949
MLE Pumping		565	565	565	565
N_2O Emissions non-BNR	947				
N_2O Emissions BNR		888	888	888	888
RAS Pumping	141	141	141	141	141
WAS Pumping	85	56	56	56	56
Sub-Total Act. Sludge	2,980	3,820	4,430	4,430	4,430
MeOH Tert. Denit. Filter				284	285
Filter Pumping			226	452	452
Alum (Filtration)			165	422	423
Conventional Filtration (Backwash)			56	56	56
Polymer (High Rate Clarifier)				72	72
High Rate Clarifier (Backwash)				40	40
Microfiltration (Pumping)					161
Microfiltration (Backwash)					40
RO (Transfer Pump)					54
RO (Feed Pump)					2,691
RO (Backwash)					40
DWI					2,476
Disinfection (Hypochlorite)	203	204	203	203	197
Sub-Total Tertiary	200	200	650	1,530	6,990
Sub-Total Liquid Stream	3,420	4,260	5,320	6,200	11,660
Solids Treatment					
GBT (Pumping)	56	56	56	56	56
GBT (Polymer)	6	5	6	6	6
Digester (Mixing)	354	354	354	354	354

Treatment	Level 1 No N/P Removal (CO_2 eq mt/yr)	Level 2 8 mg N/L 1 mg P/L (CO_2 eq mt/yr)	Level 3 4-8 mg N/L 0.1-0.3 mg P/L (CO_2 eq mt/yr)	Level 4 3 mg N/L 0.1 mg P/L (CO_2 eq mt/yr)	Level 5 <2 mg N/L <0.02 mg P/L (CO_2 eq mt/yr)
Digester (CH4)	-844	-666	-724	-656	-719
CH4 Emissions	138	109	119	107	118
Centrifuge (Polymer)	11	11	12	12	12
Centrifuge (Energy)	84	76	84	67	86
Fermentation				59	
Biosolids Hauling	48	45	50	49	52
Sub-Total Solids	-146	-10	-40	50	-30
Miscellaneous					
Miscellaneous	1,320	1,320	1,320	1,320	1,320
Sub-Total Miscellaneous	1,320	1,320	1,320	1,320	1,320
Total	**4,594**	**5,570**	**6,600**	**7,570**	**12,950**

APPENDIX B

A DETAILED DISCUSSION ON NITROUS OXIDE EMISSIONS AT WWTPS

As stated within the main body of this report, the IPCC (2006) and U.S. EPA (2009) do not adequately quantify N_2O emission factors. Both approaches assume that N_2O emissions only occur during denitrification in the anoxic zone. Although previously thought to not produce N_2O emissions, nitrification can lead to N_2O emissions. The mechanisms associated with N_2O emissions were described in Section 1.9.4.3. Despite advancements in monitoring analyses and a further understanding on the mechanisms associated with N_2O emissions at the gene level, a unified emission factor and/or algorithm for calculating N_2O emissions for wastewater treatment processes is not available at this time.

Rather than assume a particular BNR configuration for N_2O emissions and the corresponding emission factor from the U.S. nationwide survey compiled by Ahn et al. (2010b), this study assumed the average N_2O emission factors for non-BNR (three plants in total) and BNR (8 plants in total) plants in Ahn et al. (2010b). Furthermore, to account for the wide variability in the nationwide survey, the data ranges were provided as bars. The Ahn et al. (2010b) values used in this study for non-BNR (Level 1) were 32 g N/PE/yr (data range from 5 to 97 g N/PE/yr) and for BNR (Levels 2 to 5) 30 g N/PE/yr (data range from 0.3 to 140 N/PE/yr).

A discussion on potential N_2O emissions provided by Kartik Chandran is provided below.

Level 1

Based on ongoing studies at full-scale as part of the Water Environment Research Foundation (WERF) project U4R07, inefficient nitrification and denitrification have been implicated as the main factors leading to N_2O emissions from both BNR and non-BNR processes (Ahn et al., 2010a). It should be noted that plants designed just for secondary treatment could also potentially nitrify part or all of their influent total Kjeldahl nitrogen (TKN) load at high temperatures (e.g., during summer).

Under such circumstances, N_2O emissions are expected to be especially high during periods of incomplete nitrification (incomplete ammonia, NH_3 conversion to either nitrite, NO_2^- or nitrate, NO_3^-) or during periods of high NO_2^- accumulation. Incomplete nitrification is highly likely in plants not specifically designed for nitrification since these plants have a lower target operating solids retention time (SRT) than plants designed for nitrification.

Consequently, incomplete nitrification and NO_2^- accumulation is especially promoted (Ahn et al., 2010a). Yet another factor leading to N_2O emissions from nitrification is the imposition of excess influent TKN loads. Since these loads are not adequately buffered at low operating SRT, the diurnal peaks in influent TKN loading can lead to especially high NH_3 accumulation in such plants. Given that N_2O emissions from nitrification are correlated with the extent of NH_3 accumulation, this is yet another factor exacerbating N_2O emissions from plants not designed and operated for BNR, but where intermittent nitrification can occur (Ahn et al., 2010a; Yu et al., 2010).

Level 2

The emissions from an MLE configuration incorporating biological phosphorous removal without chemical addition are expected to be lower than that for just secondary treatment. This reduction is primarily because of the higher operating SRT. Owing to the higher target SRT in Level 2, the influent TKN loads are better buffered than in the case without BNR. Additionally, the standing NH_3 and NO_2^- concentrations in the MLE system are also lower than Level 2, which leads to a reduction in nitrification and denitrification associated N_2O emissions.

There are two specific strategies for optimizing the MLE configuration that could minimize the overall N_2O emissions potential, for a given effluent TN concentration of 8 mg-N/L. The first strategy is to achieve near complete nitrification, so as to minimize the standing NH_3 concentrations in the system, which implies that NO_3 will constitute the majority of the effluent TN load. The second strategy is to minimize the extent of plug-flow in the system, especially in the aerobic zone and thus minimize the NH_3 gradient therein (Ahn et al., 2010a; Yu et al., 2010). Given the positive correlation between NH_3 concentrations and N_2O emissions, minimizing plug-flow conditions can also minimize the magnitude of N_2O emissions especially in the upstream regions of the aerobic zone.

It is also conceivable that COD limitation in the anoxic zones can result in N_2O emissions as documented in some studies previously (Hanaki et al. 1992). However, through all our intensive monitoring campaigns, N_2O emissions in anoxic zones have been minimal across the board, both in the presence and absence of external COD addition (Ahn et al., 2010a). Recent lab-scale results also point to the fact that transient COD limitation contributes minimally to N_2O emissions (Lu and Chandran, 2010).

Level 3

Extending the above argument (for Level 2), the magnitude of N_2O emissions with Bardenpho for an even lower effluent TN concentration with methanol addition will be even lower. It is presumed that the lower TN concentrations will be achieved by a mix of higher SRT and external COD addition. The higher SRT will presumably stabilize nitrification, especially in response to diurnal load variations of TKN, leading to lower standing concentrations of NH_3 and NO_2^- in the system.

Again, N_2O emissions can be minimized for a given effluent TN concentration, by minimizing the effluent NH_3 fraction thereof. Further reductions in N_2O emissions can also be effected by minimizing the plug-flow regime, especially of the primary aerobic zone (Ahn et al., 2010a). It should be pointed out that the differences in the N_2O emissions between Level 2 and Level 3 cannot be directly correlated to differences in the configuration (MLE *vs* Bardenpho) alone, since different levels of TN removal are being achieved in the two configurations.

Rather, the different levels of N_2O emissions are due to the different operating SRTs and the correspondingly different nitrogen concentrations and speciation therein. Due to the low magnitude of N_2O emissions from denitrification observed consistently to date, differences in effluent NO_3^- concentrations between Level 2 and Level 3 are not expected to contribute significantly to differences in N_2O emissions (Ahn et al., 2010a).

Level 4

N_2O emissions in Level 4 are expected to be similar or minimally lower than for Level 3, if and only if the main difference is in the effluent nitrate concentrations and not in effluent NH_3 concentrations. There is always the danger of overdosing external COD such that heterotrophic growth occurs at the expense of growth of nitrifying bacteria and significant N-removal is by virtue of N-assimilation (Grady et al., 1999). In such a unique scenario, nitrification could be intermittent and unstable (similar to Level 2), resulting in N_2O emissions. However, this unique scenario is expected to be highly unlikely and avoidable given the use of process simulation tools for design and optimization.

Level 5

Assuming that the only advantage of Level 5 over Level 4 is the removal of additional soluble organics, the N_2O emissions in both Levels 4 and 5 are expected to be comparable.

Impact of Different Levels of Enhanced Biological Phosphorus Removal on N_2O Emissions

Finally, although there is some evidence on the impact of enhanced biological phosphorus removal (EBPR) on N_2O emissions, the specific mechanisms are related to transient depressions in pH rather than specific EBPR reactions and pathways themselves (Zhou et al., 2008). Thus, the different levels of EBPR are not expected to contribute differentially to N_2O emissions from the different systems considered herein.

Summary

Based on the different levels of BNR treatment and corresponding operating SRTs, N_2O emissions are expected in the following order: Level 1 > Level 2>Level 3 > (Level 4~Level 5). Higher standing concentrations of NH_3 and NO_2 primarily influence the magnitude of N_2O emissions from the different configurations. Strategies and configurations aimed at minimizing these concentrations (both at steady-state and diurnal transients) will consequently lead to lower N_2O emissions.

The impact of factors such as non-optimal aeration and oxygen intrusion into anoxic zones is not explicitly considered here, since these factors ultimately result in build-up of NH_3, and NO_2 among other N-species, which in turn influence N_2O emissions. For a more detailed analysis, the impacts of temperature will also need to be considered for their direct and indirect influence on N_2O emissions. The development of process models to perform a more detailed evaluation of these and additional factors is now underway as part of WERF U4R07.

REFERENCES

Ahn, J.-H., Kim, S., Pagilla, K., Katehis, D., Chandran, K. (2010a) Spatial and temporal variability in N_2O generation and emission from full-scale BNR and non-BNR processes. *Wat. Environ. Res.*, 82(12):2362-2372.

Ahn, J.-H., Kim, S, Park, H., Rahm, B., Pagilla, K., Chandran, K. (2010b) N_2O emissions from activated sludge processes, 2008-2009: Results of a national surveying program in the United States. *Environ. Sci. Technol.*, 44(12):4505-4511.

APHA, American Public Health Association, American Water Works Association, and the Water Environment Federation (2005) *Standard Methods for the Examination of Water and Wastewater, 21st Edition.* Washington, D.C.

Ashley, R., Blackwood, D., Butler, D., Jowitt, P., Davies, J., Smith, H., Gilmour, D., Oltean-Dumbrava, C. (2008) Making asset investment decisions for wastewater systems that include sustainability. *J. Environ. Engr. – ASCE*, 134(3):200-209.

Baytshtok, V., Kim, S., Yu, R., Park, H., Chandran, K. (2008) Molecular and biokinetic characterization of methylotrophic denitrification using nitrate and nitrite as terminal electron acceptors. *Wat. Sci. Technol.*, 58(2):359-365.

Baytshtok, V., Lu, H., Park, H., Kim, S., Yu, R., Chandran, K. (2009) Impact of varying electron donors on the molecular microbial ecology and biokinetics of methylotrophic denitrifying bacteria. *Biotechnol. Bioengr.*, 102(6):1527-1536

Beck, M.B., et al. (In press) Technology, sustainability, and business: cities as forces of good in the environment. In *Integrated Urban Water Management in Temperate Climates*; Maksimovic, C., Ed.; UNESCO and Taylor & Francis: London.

Biswas, W.K. (2009) Life cycle assessment of seawater desalinization in Western Australia. *World Acad. Sci, Engr. Technol.*, 56:369-275.

Bock, E. (1995) Nitrogen loss caused by denitrifying Nitrosomonas cells using ammonium or hydrogen as electron donors and nitrite as electron acceptor. *Arch. Microbiol.*, 163:16-20.

Bott, C. and Parker, D. (2011) WEF/WERF Study Quantifying Nutrient Removal Technology Performance. WERF Research Project under Nutrient Challenge, NUTR1R06h.

Bourgeous, K., B. Narayanan, B., and G. Deis (2003). Particle size distribution testing as a diagnostic tool for optimizing filtration and UV disinfection systems. Proceedings of 76th Annual Water Environmental Federation Annual Conference and Exposition, Wastewater Treatment Research, Los Angeles, CA, October, 2003.

Butler, M.D., Wang, Y.Y., Cartmell, E., Stephenson, T. (2009) Nitrous oxide emissions for early warning of biological nitrification failure in activated sludge. *Wat. Res.*, 43:1265-1272.

California Air Resources Board, ICLEI, California Climate Action Registry, The Climate Registry (2008) Local Government Operations Protocol. For the quantification and reporting of greenhouse gas emissions inventories, Version 1.1.

Cakir, F.Y. and Stenstrom, M.K. (2005) Greenhouse gas production: A comparison between aerobic and anaerobic wastewater treatment technology. *Wat. Res.*, 39:4197-4203.

Chan, L., Leu, S.Y., Rosso, D., Stenstrom, M.K. (In press) The relationship between mixed-liquor particle size and solids retention time in the activated sludge process. *Wat. Environ. Res.*

Chandran, K. et al. (2010) Greenhouse nitrogen emission from wastewater treatment operations: Interim report, WERF research project U4R07a.

Clark, D., Hunt, G., Kasch, M.S., Lemonds, P.J., Moen, G.M., Neethling, J.B. (2010) Nutrient management: Regulatory approaches to protect water quality. Volume 1 – Review of existing practices, WERF research project NUTR1R06i.

Coke, L.B., Weir, C.C., Hill, V.G. (1987) Environmental impact of bauxite mining and processing in Jamaica. *Social and Economic Studies*, 36:289-333.

Conley, D.J., Paerl, H.W., Howarth, R.W., Boesch, D.F., Seitzinger, S.P., Havens, K.E., Lancelot, C., Likens, G.E. (2009) Controllling eutrophication: nitrogen and phosphorus. *Sci.*, 323(5917):1014-1015.

Czepiel, P., Crill, Pl, Harriss, R. (1995) Nitrous oxide emissions from municipal wastewater treatment. *Environ. Sci. Technol.*, 29:2352-2356.

Daigger, G.T. (2009) Evolving Urban Water and Residuals Management Paradigms: Water Reclamation and Reuse, Decentralization, and Resource Recovery. *Wat. Environ. Res.*, 81(8):809-823.

de Haas, D.W., Hartley, K.J. (2004) Greenhouse Gas Emissions from BNR Plant-Do We Have the Right Focus? In: Proceedings of the Sewage Management-Risk Assessment and Triple Bottom Line. Queensland Environmental Protection Agency, Cairns, Australia, April 5-7.

Dimitriou-Christidis, P. and Wilner, H.T. (2009) Gaseous emissions from wastewater facilities. *Wat. Environ. Res.*, 81(10):1394-1405.

Doyle, J.D., Parson, S.A. (2002) Struvite formation, control and recovery. *Wat. Res.* (36):3925-3940.

Drury, D. Shepherd, W., Narayanan, B. (2005) Phosphorus – How low can you go? Proceedings of 78th Annual Water Environmental Federation Annual Conference and Exposition, Wastewater Treatment Research, Washington, D.C.

Dwyer, J., Kavanagh, L., Lant, P. (2008a) The degradation of dissolved organic nitrogen associated with melanoidin using a UV/H2O2 AOP. *Chemosphere*, 71(9):1745-1753.

Dwyer, J. & Lant, P. (2008b) Biodegradability of DOC and DON for UV/H2O2 pre-treated melanoidin based wastewater. *J. Biochem. Engr.*, 42(1):47-54.

EPA (2010) Total Maximum Daily Load (TMDL) Program. EPA Region 4, Atlanta, GA.

Esparza-Soto, M., Lee, W.M., Mash, H., Westerhoff, P. (2003) Dissolved organic nitrogen removal during coagulation and PAC treatments. Proceedings of the Annual Meeting of AWWA, 486-496.

Finch, R. (2009) Presentation at the Sustainable Future Preconference Workshop of the Annual Pacific Northwest Clean Water Association Conference, Boise, ID. http://www.pncwa.org/assets/documents/2009%20PNCWA%20Sustainable%20Future%20Workshop%20-%20Robbin%20Finch.pdf

Foley, J., de Haas, D., Hartley, K., Lant, P. (2010) Comprehensive life cycle inventories of alternative wastewater treatment systems. *Wat. Res.*, 44:1654-1666.

Girsch, P., De Vries, S. (1997) Purification and initial kinetic and spectroscopic characterization of NO reductase from *Parococcus denitrificans*. *Biochimica et Biophysica Acta (BBA) – Bioenergetics*, 1318:202-216.

Goreau, T.J., Kaplan, W.A., Wofsy, S.C., McElroy, M.B., Valois, F.W., Watson, S.W. (1980) Production of NO_2^- and N_2O by nitrifying bacteria at reduced concentrations of oxygen. *Appl. Environ. Microbiol.*, 40(3):526-532.

Gu, A.Z., Saunders, A., Neethling, J.B., Stensel, H.D., Blackall, L.L. (2008) Functionally relevant microorganisms to enhanced biological phosphorus removal performance at full-scale wastewater treatment plants in the United States. Wat. Environ. Res., 80(8):688-698.

Gu, A.Z., et al. (2010) Protocol to evaluate alternative external carbon sources for denitrification at full-scale wastewater treatment plants, WERF research project NUTR1R06b.

Grady, C. P. L. J., G. T. Daigger, and H. C. Lim (1999) *Biological Wastewater Treatment, 2nd Ed.* Marcel Dekker, New York.

Guest, J.S., Skerlos, S.J., Daigger, G.T., Corbett, J.R.E., Love, N.G. (2010) The use of qualitative system dynamics to identify sustainability characteristics of decentralized wastewater management alternatives. *Wat. Sci. Technol.*, 61(6):1637-1644.

Guest, J.S., Skerlos, S.J., Barnard, J.L., Beck, M.B., Daigger, G.T., Hilger, H., Jackson, S.J., Karvazy, K., Kelly, L., Macpherson, L., Mihelcic, J.R., Pramanik, A., Raskin, L., van Loosdrecht, M.C.M., Yeh, D., Love, N.G. (2009) A new planning and design paradigm to achieve sustainable resource recovery from wastewater. *Environ. Sci. Technol.*, 43:6126-6130.

Guisasola, A., de Haas, D., Keller, J., Yuan, Z. (2008) Methane formation in sewer systems. *Wat. Res.*, 42:1421-1430.

Guisasola, A., Sharma, K.R., Keller, J., Yuan, Z. (2009) Development of a model for assessing methane formation in rising main sewers. *Wat. Res.*, 43:2874-2884.

Gur-Reznik, S., Katz, I., Dosoretz, C. (2008) Removal of dissolved organic matter by granular-activated carbon adsorption as a pretreatment to reverse osmosis of membrane bioreactor effluents. *Wat. Res.*, 42:1595-1605.

Hanaki, K., Hong, Z., Matsu, T. (1992) Production of nitrous oxide gas during denitrification of wastewater. *Wat. Sci. Technol.*, 26:1027-1036.

Ho, C.-F. H., Jolis, D., Tansel, B. (2007) Gaseous emissions from wastewater facilities. *Wat. Environ. Res.*, 79(10):1528-1550.

Hooper, A.B. (1969) Lag phase of ammonia oxidation by resting cells of *Nitrosomonas europaea*. *J. Bacteriol.*, 97:968-969.

Horz, H.-P., Barbrook, A., Field, C.B., Bohannan, B.J.M. (2004) Ammonia-oxidizing bacteria respond to multifactorial global change. *Proceedings of the Nat. Acad. Sci.*, 101(42):15136-15141.

Hunkeler, D. (2006) Societal LCA Methodology and Case Study. *Int. J. Life Cycle Assess.*, 11(6):371-382.

Hunt, P., TzeYen, C., Franklin, B., Vermande, S., Wilson, T. (2009) A critical new look at nutrient control processes-2009 update. Proceedings of 82nd Annual Water Environmental Federation Annual Conference and Exposition, Wastewater Treatment Research, Orlando, FL.

Insam, H. and Wett, B. (2008) Control of GHG emission at the microbial community level. *Waste Mgmt.*, 28:699-706.

Intergovernmental Panel on Climate Change (2000) Special Report on Emissions Scenarios: A Special Report of Working Group III of the Intergovernmental Panel on Climate Change. Cambridge University Press, Cambridge, UK.

Intergovernmental Panel on Climate Change (2006) *2006 IPCC Guidelines for National Greenhouse Gas Inventories. Prepared by the National Greenhouse Gas Inventories Programme*, Eggleston, S., Buendia, L., Miwa, K., Ngara, T., Tanabe, K. (eds.) Published: IGES, Japan.

Jenkins, D., Richard, M.G., Daigger, G.T. (2004) Manual on the causes and control of activated sludge bulking, foaming, and other solids separation problems, (3rd Edition). IWA publishing, CRC Press, London, UK.

Kampschreur, M.J., Tan, N.C.G., Kleerebezem, R., Picioreanu, C., Jetten, M.S.M., Loosdrecht, M.C.M.V. (2008a) Effect of dynamic process conditions on nitrogen oxides emission from a nitrifying culture. Environ. Sci. Technol., 42:429-435.

Kampschreur, M.J., Van Der Star, W.R.L., Wielders, H.A., Mulder, J.W., Jetten, M.S.M., van Loosdrecht, M.C.M. (2008b) Dynamics of nitric oxide and nitrous oxide emission during full-scale reject water treatment. *Wat. Res.*, 42:812-826.

Kampschreur, M.J., Temmink, H., Kleerebezem, R. Jetten, M.J.M., van Loosdrecht, M.C.M. (2009) Nitrous oxides emission during wastewater treatment. *Wat. Res.*, 43(17):4093-4103.

Kang, S.J., Olmstead, K.P., Takacs, K.M, Collins, J., Wheeler, J., Zharaddine, P. (2009) Energy sustainability and nutrient removal from municipal wastewater. Proceedings of 82nd Annual Water Environmental Federation Annual Conference and Exposition, Wastewater Treatment Research, Orlando, FL.

Keller, J.V., Leckie, J.O., and McCarty, P.L. (1978) Investigation of soluble organic nitrogen compounds in municipal secondary effluent. *J. Wat. Pollution Control Fed.*, 50(11):2522-2529.

Keller, J. and Hartley, K. (2002) Greenhouse gas production in wastewater treatment: process selection is the major factor. *Wat. Sci. Technol.*, 47(12):43-48.

Kester, R.A., de Boer, W., Laanbroek, H.J., (1997) Production of NO and N_2O by pure cultures of nitrifying and denitrifying bacteria during changes in aeration. *Appl Environ. Microbiol.*, 63:3872-3877.

Khan, E. (2007) Development of a technology based biodegradable dissolved organic nitrogen (BDON) protocol. *Presentation at Water Environment Research Foundation and Chesapeake Bay Science and Technology Advisory Committee Workshop*, Baltimore, MD, September 27/28.

Kimochi, Y., Inamori, Y., Mizuochi, M., Xu, K.-Q., Matsumura, M. (1998) Nitrogen removal and N_2O emission in a full-scale domestic wastewater treatment plant with intermittent aeration. *J. Ferment. Bioengr.*, 86:202-206.

Kolpin, D.W., Furlong, E.T., Meyer, M.T., Thurman, E.M.; Zaugg, S.D.; Barber, L.B.; Buxton, H.T. (2002) Pharmaceuticals, hormones, and other organic wastewater contaminants in U.S. streams, 1999-2000: a national reconnaissance. *Environ. Sci. Technol.*, 36(6):1202-1211.

Larsen, T.V., Alder, A.C., Eggen, R.I.L., Maurer, M., Lienert, J. (2009) Source separation: Will we see a paradigm shift in wastewater handling? *Environ. Sci. Technol.*, 43(16):6121-6125.

Leu, S.-Y., Chan, L., Stenstrom, M.K. (In press) Toward long SRT of activated sludge processes: Benefits in energy saving, effluent quality, and stability. *Wat. Environ. Res.*

Li, Q., Li, P., Zhu, P., Wu, J., Liang, S. (2008) Effects of exogenous organic carbon substrate on nitrous oxide emissions during denitrification process of sequential batch reactors. *Environ. Engr. Sci.*, 25(8):1221-1228.

Liao, B.Q., Allen, D.G., Droppo, I.G., Leppard, G.G., Liss, S.N. (2001) Surface properties of sludge and their role in bioflocculation and settleability. *Wat. Res.*, 35:339-350.

Liu, L. and Gu, A.Z. (2010) Phosphorus speciation and removal in wastewater treatment: Mid-point result summary report. Water Environment Research Foundation, Alexandria, VA.

Lu, H. and K. Chandran (2010a) Factors promoting emissions of nitrous oxide and nitric oxide from denitrifying sequencing batch reactors operated with methanol and ethanol as electron donors. *Biotechnol. Bioeng.*, 106(3):390-398.

Lu, H. and K. Chandran (2010b) Diagnosis and quantification of glycerol assimilating denitrifying bacteria in an integrated fixed-film activated sludge reactor via C-13 DNA stable-isotope probing. *Environ. Sci. Technol.*, 44(23):8943-8949.

Lundie, S., Peters, G.M., Beavis, P.C. (2004) Life cycle assessment for sustainable metropolitan water systems planning. *Environ. Sci. Technol.*, 38(13):3465-3473.

McKinney, R. E. (1962). Mathematics of complete mixing activated sludge. *J. Sanitary Engineering Division, ASCE*, 88:87-113.

MDEQ (2001) Draft nutrient management plan and total maximum daily load for Flathead Lake. Montana Department of Environmental Quality, Helena, MT.

Mickley and Associates (2006) *Membrane Concentrate Disposal: Practices and Regulations*. U.S. Department of the Interior, Bureau of Reclamation, Contract No. 98-FC-81-0054.

Minami, K., Takata, K. (1997) Atmospheric methane: sources, sinks and strategies for reducing agricultural emissions. *Wat. Sci. Technol.*, 36(6-7):509-516.

Monteith, H.D., Sahely, H.R., MacLean, H.L., Bagley, D.M. (2005) A rationale procedure for estimation of green-house gas emissions from municipal wastewater treatment plants. *Wat. Environ. Res.*, 77(4):390-403.

National Biosolids Partnership (NBP) (2007) 2006-2007 Annual report – environmental stewardship: a decade of progress in implementing a national model EMS program for the water quality profession. Alexandria, VA, www.biosolids.org/docs/NBP_2006_2007_AnnualReport.pdf.

Neethling, J.B., Bakke, B., Benisch, M., Gu, A.Z., Stephens, H., Stensel, H.D., Moore, R. (2005) Factors influencing the reliability of enhanced biological phosphorus removal. , WERF research project 01-CTS-3.

Neethling, J.B., Benisch, M., Clark, D., Fisher, D., Gu, A.Z. (2007) Phosphorus speciation provides direction to produce 10 µg/L. Proceedings of 80th Annual Water Environmental Federation Annual Conference and Exposition, Wastewater Treatment Research, San Diego, October, 2007.

Palm, J.C., Jenkins, D., Parker, D. (1980) Relationship between organic loading, dissolved oxygen concentration and sludge settleability in the completely-mixed activated sludge process, *J. Wat. Pollution Control Fed.*, 52(10):2483-2506.

Oki, T., Kanae, S. (2006) Global hydrological cycle and world water resources. *Science*, 313(5790): 1068-1072.

Owen, W.F. (1982) Energy in Wastewater Treatment. Prentice-Hall, Englewood Cliffs, New Jersey.

Pagilla, K. (2007) Presentation at Water Environment Research Foundation and Chesapeake and Bay Science and Technology Advisory Committee Workshop, Baltimore, MD, September 27/28.

Pagilla, K., Shaw, A., Kunetz, T., Schiltz, M. (2009) A systematic approach to establishing carbon footprints for wastewater treatment plants. Proceedings of 82nd Annual Water Environmental Federation Annual Conference and Exposition, Wastewater Treatment Research, Orlando, FL.

Pehlivanoglu, E. & Sedlak, D.L. (2004) Bioavailability of wastewater-derived organic nitrogen to the alga Selenastrum Capricornutum. *Wat. Res.*, 38(14-15):3189-3196.

Préndez , M. and Lara-González (2008) Application of strategies for sanitation management in wastewater treatment plants in order to control/reduce greenhouse gas emissions, *J. Environ. Mgmt.*, 88:658-664.

Racoviceanu, A.I., Karney, B.W., Kennedy, C.A., Colombo, A.F. (2007) Life-cycle energy use and greenhouse gas emission inventory for water treatment systems. *J. Infrastructure Systems ASCE*, 13(4):261-270.

Rogers, P.; Silva, R.D.; Bhatia, R. (2002) Water is an economic good: How to use prices to promote equity, efficiency, and sustainability. *Wat. Policy*, 4:1-17.

Sandinho, J. et al. (2008) Carbon Augmentation for Biological Nutrient Removal. WERF nutrient removal challenge compendium.

Schmidt, I., Bock, E. (1997) Anaerobic ammonia oxidation with nitrogen dioxide by *Nitrosomonas eutropha*. *Arch. Microbiol.*, 167:106-111.

Schmidt, I., Bock, E. (1998) Anaerobic ammonia oxidation by cell-free extracts of *Nitrosomonas eutropha*. *Antonie van Leeuwenhoek*, 73:271-2781

Shahabadi, M.B., Yerushalmi, L., Haghighat, F. (2009) Impact of design on greenhouse gas (GHG) generation by wastewater treatment plants. *Wat. Res.*, 43:26792687.

SimaPro 6 (2008) Life Cycle Analysis Software. The Netherlands.

Smith, D. S., Takács, I., Murthy, S., Daigger, G.T., and Szabo, A. (2008) Phosphate complexation model and its implications for chemical phosphorous removal. *Wat Environ. Res.*, 80(5):428-438.

Snoeyink, V. & Jenkins, D (1980) *Water Chemistry*. John Wiley & Sons. New York, NY.

Sommer, J., Ciplak, A., Sumer, E., Benckiser, G., Ottow, J.C.G. (1998) Quantification of emitted and retained N_2O in a municipal wastewater treatment plant with activated sludge and nitrification-denitrification units. *Agrobiol. Res.*, 51:59-73.

Stasinakis, A.S., Thomaidis, N.S., Nikolaos, S., Mamais, D., Karivali, M. Lekkas, T.D. (2003) Chromium species behavior in the activated sludge process. Chemosphere, 52(6):1059-1067.

Stensel, H.D. et al. (2008) Dissolved organic nitrogen (DON) in biological nutrient removal wastewater treatment processes. WERF research compendium under Nutrient Challenge, NUTR1R06.

Sümer, E., Weiske, A., Benckiser, G., Ottow, J.C.G. (1995) Influence of environmental conditions on the amount of N_2O released from activated sludge in a domestic waste water treatment plant. *Cell. Mol. Life Sci.*, 51:419-422.

Tallec, G., Garnier, J., Billen, G., Gousailles, M. (2006) Nitrous oxide emissions from secondary activated sludge in nitrifying conditions of urban wastewater treatment plants: Effect of oxygenation level. *Wat. Res.*, 40:2972-2980.

Tallec, G., Garnier, J., Billen, G., Gousailles, M. (2008) Nitrous oxide emissions from denitrifying activated sludge of urban wastewater treatment plants, under anoxia and low oxygenation. *Biores. Technol.*, 99:2200-2209.

The Climate Registry (2008) *General Reporting Protocol, Version 1.1 – Acute, transparent, and consistent measurement of greenhouse gases across North America*.

Tripathi, M. (2007) Life-Cycle Energy Emissions for Municipal Water and Wastewater Services: Case-Studies of Treatment Plants in US. Master's Thesis from the U. of Michigan, Ann Arbor, MI.

U.S. EPA (1995) *Human Health Benefits from Sulfate Reduction Under Title IV of the 1990 Clean Air Act Amendments. EPA-430-R-95-010.* Washington, D.C: U.S. Environmental Protection Agency.

U.S. EPA (2007) *The Emissions & Generation Resource Integrated Database – eGrid WebVersion1.0.* United States Environmental Protection Agency, Washington, D.C.

U.S. EPA (2009) Inventory of U.S. Greenhouse Gas Emissions and Sinks: 1990-2006, EPA 430-R-08-005. Washington, D.C.

U.S. EPA (2010) Water quality standards for the State of Florida's lakes and flowing waters; proposed rule (Part III). 40 CFR Part 131.

Vader, J.S., van Ginkel, C.G., Sperling, F., de Jong, J., de Boer, W., de Graaf, J. S., van der Most, M., Stokman, P.G.W. (2000) Degradation of ethinyl estradiol by nitrifying activated sludge. *Chemosphere* 41(8): 1239-1243.

Vallentyne, J.R. (1974) The algal bowl – Lakes and man. Misc. Special Publ. 22, Dept. of the Environment, Ottawa. 185pp.

Wallis-Lage, C. Barnard, J.L., deBardadillo, C. (2009) Perspectives in energy use and greenhouse gases at wastewater treatment plants. Proceedings of the 82nd Annual Water Environment Federation Technical Exposition and Conference, Orlando, Florida.

Water Environment Federation (2009). *Design of Municipal Wastewater Treatment Plants, WEF Manual of Practice 8, Fourth Edition, ASCE Manuals and Reports on Engineering Practice No. 76, Volume 1.* Alexandria, VA.

Water Environment Federation (2010) *Biological and Chemical Systems for Nutrient Removal, Special Publication.* Water Environment Federation: Alexandria, Virginia.

Westerhoff, P., Esparza-Soto, M. (2003) Occurrence and treatability of DON in water supplies. Proceedings of the Annual Meeting of AWWA, 1120-1123.

Wett, B., Buchauer, K. Fimmi, C. (2007) Energy self-sufficiency as a feasible concept for wastewater treatment systems. Proceedings of the International Water Association Leading Edge Conference.

Wicht, H., Beier, M. (1995) N_2O emissions aus nitrifizierenden un denitrificierended Klaranlagen. *Korrespondez Abwasser*, 42:404-406, 411-413.

Williams, J., Larocque, S. (2009) Calculating a sustainable return on investment. *J. Public Works & Infrastructure*, 2(2):94-105.

Willis, J., Bailey, W., Jolly, E., Murthy, S., Peot, C., Trueblood, D. (2009) DCWASA's certifiable, utility-wide GHG inventory. Proceedings of the 82nd Annual Water Environment Federation Technical Exposition and Conference, Orlando, Florida.

Wolstenholme, E.F. (1999) Qualitative vs Quantitative Modelling: The Evolving Balance. *J. Oper. Res. Soc.*, 50(4):422-428.

World Commission on Environment and Development (1987) Our Common Future, Report of the World Commission on Environment and Development. Published as Annex to General Assembly document A/42/427.

Xu, B., Li, D.-P., Li, W., Xia, S.-J., Lin, Y.-L., Hu, C.-Y., Zhang, C.-J., Gao, N.-Y. (2010) Measurements of dissolved organic nitrogen (DON) in water samples with nanofiltration pretreatment. *Wat. Res.*, 44:5376-5384.

Yu, R., Kampschreur, M., van Loosdrecht, M.C.M., Chandran, K. (2010) Mechanisms and specific directionality in autotrophic nitrous oxide and nitric oxide generation during transient anoxia. *Environ. Sci. Technol.*, 44(4):1313-1319.

Zhou, Y., Pijuan, M., Zeng, R. J., Yuan, Z. (2008) Free nitrous acid inhibition on nitrous oxide reduction by a denitrifying-enhanced biological phosphorus removal sludge. *Environ. Sci. Technol.*, 42:8260-8265.

Zimmerman, J.B., Mihelcic, J.R., Smith, J. (2008) Global stressors on water quantity and quality. *Environ. Sci. Technol.*, 42(12):4247-4254.

IWA Publishing's authorised EU representative for General Product Safety Regulations is Diane D'Arras, 15 rue Duret, 75116 Paris, France, e-mail: safety@iwap.co.uk.

Printed and bound by CPI Group (UK) Ltd, Croydon, CR0 4YY
12/05/2026
02108881-0011